Successful Harvesting of Nature's Free Foods

Includes: Gardening, hiking, camping, survival, and general outdoor tips.

Ken Larson

Illustrations and Photography by Ken Larson

Wild Foods of United States, Britain, and Northern Europe

SECOND EDITION

RHEMA PUBLISHING INC.
P.O. Box 789 Suwanee, GA 30024
770-932-6991

http://members.aol.com/keninga/index.htm

CAUTION: In the identification and use of wild edibles as a food source, care and attention to details should be exercised, as some plants are toxic. Always use several field guides to insure proper identification. Before eating any wild plants you collect, you should have been field trained by an expert. If this is not possible, take the plant to an expert or botanist for positive identification. Even after these precautions, initially taste only a small amount of the plant in question. You should then wait several hours before tasting more. After all, enjoyment is the goal!

GOD'S FREE HARVEST

 Publisher: Rhema Publishing, Inc., P.O. Box 789, Suwanee, GA 30024.

To reorder: Write Rhema Publishing, Inc. P.O. Box 789, Suwanee, GA 30024. Phone: 770-932-6991. Cost $12.95. Postage and Handling: First book $2.00. Additional books $1.00 each.

Library of Congress Cataloging and Publication Data

Larson, Ken
God's Free Harvest
Includes index.
1. Gardening 2. Wild plants, Edible - Identification.
3. Camping 4. Survival

ISBN 0-9642497-0-7

CHAPTER I

INTRODUCTION TO WILD FOODS

PRECAUTIONS TO OBSERVE

There are uses–or precautions to observe–of all parts of a plant and in all seasons. For example, cattail roots can be used at any time of the year, but they're best from late fall through early spring, when the starch concentration is higher. (After spring the roots shrink and harden.) They can be baked, or the starch can be processed out of them.

The corms, or spouts, are gathered from late summer to winter. These can be cooked like potatoes. Shoots, gathered in spring, can be steamed or boiled and taste like cucumbers. Pollen, produced in early spring, can be eaten raw, boiled in water until it thickens into a hot cereal, or use as a protein booster in breads or soups. In late spring the flower spikes are gathered and boiled, with a flavor like corn. Seeds collected in summer are mashed into a protein-rich flour.

LIFE SAVING KNOWLEDGE

A knowledge of edible wild plants and herbs can be interesting, giving the student both new culinary experiences and a greater appreciation of the world around us. In extreme situations, such knowledge could be extremely valuable, or even life-saving.

As a disaster specialist with the Federal Government, I regularly see instances of our food system collapsed and our unprepared population reacting in panic to a shortage situation. During recent disasters, I have seen fights break out in grocery stores between good people over water for sale at $10.00 a gallon, ice at $15.00 a bag, and candles marked up 500 percent. Much of the need for these goods and the hardships stemming from the deprivation of necessities were direct effects of the lack of preparedness and ignorance of available resources.

Fortunately, many of the basic necessities of life are available for us; however, an alarming proportion of our people aren't aware of the simplest of these provisions. Now rather than depleting our resources, we can utilize nature's free harvest–wild plants.

The use of wild foods can reduce dependence on the current fragile food system. In addition, wild foods can be stored and, because of their wild nature, they are not depleted as easily as a conventional garden.

Wild plant gardening and foraging are invaluable skills. With it one might collect a delicious, satisfying meal and wild plants are natural and organic. They grow unattended, and offer a free, nutritious meal.

The value of the free harvest that is available to us with foraging skills provides everyday needs and may be seen in two examples:

Pine Needles

Vitamin C abounds in your yard and the wild. A palatable tea made from green pine needles contains more vitamin C than orange juice. To prepare it, cover chopped green pine needles with boiling water and let them steep for 15 minutes; then mix the liquid with sugar or use it in a soup to mask the slightly resinous taste.

Acorns

Why use animal protein? Acorns can be processed to produce an alternate protein source if they are shelled and boiled in several changes of water to reduce bitterness. The result is a good usable, tasteless protein and calorie source with the consistency of a mushroom.

WEEDS OR FOOD

Unfortunately, most people are hesitant to use wild edibles because they feel wild plants are "weeds." They feel wild edibles are just weeds not as good as food plants from the produce section of the grocery store and they fear being poisoned by a misidentified plant. Americans actually use only a small portion of the plants that are available even though they often are in demand by other cultures.

An available government publication (*USDA Composition of Foods–Vegetables*, Volume 8-11) lists many wild foods as edible, including amaranth, arrowhead, burdock, chicory, dandelion, dock, and lamb's-quarters.

Actually, most wild edibles are easy to identify while even our own "safe" garden plants can be dangerously used. For example, the green skin that forms on an uncovered potato exposed to the sun can be toxic, and the leaves of potatoes are also toxic. And what about the pesticides used by many producers?

In foraging, knowledge and experience can remove most fear. Also, many find that their initial trips to "graze" develop into an entertaining family ritual. Wild edibles can provide a practical food source for outdoorsmen, gardeners, and even city dwellers a practical food source and can even be deemed a "sport" as this book will explain.

WILD EDIBLE GARDENING

Cultivating selected wild edibles in my garden has provided me a source of great delight. Not only does adding the wild plants to my diet improve its nutritional quality, but they are also usually insect-proof, drought-resistant, and more prone to thrive in less-than-perfect soils than are the customary garden crops. In addition, cultivated wild edibles normally have larger leaves, fruits, and tubers and are milder in taste than those found in the wild.

Since they are free for the taking, wild edible gardening can add variety and pesticide-free nutrients to the diet.

Also, wild plants can be used to determine soil condition. When you see sorrel and dock in your garden, you have acid soil and it is past time to lime your soil.

In congested areas where a garden is limited, one can be very creative in picking a gardening spot and protecting the harvest. For example, most people would not recognize a potato vine if it were not in a garden row. Why not experiment with planting several potatoes in a corner of your property or in a vacant lot? A hole can be dug and filled with good soil, then potatoes planted and mulched. By the time the plant is ready to harvest, it will look like just another weed. K.L.

HOW TO PREPARE WILD FOODS

Whenever possible, wild edibles should be eaten raw for maximum nutritional value. When cooking is necessary, use a minimum amount of water, put a lid on the container, and cook as rapidly as possible to preserve nutrients.

GREENS

Using wild "greens" may be the best way to introduce wild edibles into the diet. The ones discussed here can be substituted in almost any recipe calling for garden greens. Several greens that I think especially good are lamb's-quarters, sheep sorrel, dandelion, pokeweed, and wild mustard. Of course, you can substitute wild edibles for the more common garden vegetables! For example, you could use amaranth, lamb's-quarters, or wild mustard to replace spinach. Experimentation will establish your preferences.

Normally, a young plant's leaves are less bitter than those of a mature one. To remove bitterness, however, one may usually do a quick parboiling and discard the liquid to lessen the strong taste of older wild edibles.

I like to mix a tangy fresh salad using bitter-tasting dandelion with mild lamb's-quarters or chick-weed. Mixing the greens can tone down strong tastes and add more zip to a bland one. Another favorite salad combines equal parts of several varieties of greens such as dandelion, shepherd's purse, dock, peppergrass, or sorrel. A chopped wild onion can be added to the raw greens and then a favorite dressing.

"Wilted" greens is always a favorite dish! To prepare, take several slices of fried bacon and break into small pieces. Mix the bacon and a combination of chopped dandelion, wild lettuce, or other greens. Heat three tablespoons of vinegar with bacon drippings, a dash of salt, a little sugar, and pour it over the greens and toss.

Many feel that steamed lamb's-quarters tastes better than spinach. Moreover, lamb's-quarters has a longer growing season than spinach, and it is free. Another favorite method is to cook amaranth or pokeweed (after parboiled and drained) in with scrambled eggs, as some people do spinach.

Greens are especially fun to prepare with best results early in the spring while the leaves are tender. The more you experiment, the more you will find ways to substitute wild edibles for domestic greens.

TUBERS

The roots and tubers of many plants may be used as well as the leaves for food. In order to be digestible, most roots and tubers must be cooked. Even the common potato has little nutritional value until it is cooked.

SEEDS

Seeds of many wild edibles extend the use of the plant for additional nutritional value. As a wild edible's leaves pass their season to be used as a fresh food, they produce seed.

Maple tree seeds are small, but they can be ground into flour.

Many wild edible seeds are large enough to be made into flour which typically offers a good protein supplement. To make flour, the mature seed stalk must be harvested when the plant is dry but before the seeds fall. Lay the seed stalks in the sun or near a stove to fully dry. Then, rub the seed and stalk between your hands to release the seeds.

Another method is to cut the plant off at ground level and place only the seed heads in a paper bag; allow them to dry for several weeks, and then shake the bag to separate the seed.

Unfortunately, the smaller the seed, the more difficult the separation process. It might be practical to grind and use very small seeds with the chaff.

Any of several methods are used to separate the seeds from the chaff. One method is to toss the seeds in the air, allowing a breeze to blow the chaff away while the seed falls back into a basket. Or, you may choose to slowly pour the seeds, chaff, and small pieces from one container to another in front of a fan creating air movement to blow away the chaff. I have even had luck with shaking a window screen to separate very small seeds from the chaff as women in more primitive societies have done with flat baskets. Most of the seeds fall through while the chaff remains. Another method is to shake the chaff and seed in a deep container causing the denser seeds to settle to the bottom while the hulls rise to the top. The last method is to shake the mixture in water causing the heavier seeds to sink and the hulls to float.

Try seeds as cooked cereal. Cook them with water stirring until they thicken into a mush. Add herbs and roots such as sassafras to the cooking process for variety. If a cereal is not of interest, seeds can be toasted in the oven for grinding into flour or sprinkling onto cereal or breads for texture and protein.

CANNING/FREEZING

We can counter the short growing season of some wild edibles by preserving them. Stored edibles can also add excitement to a winter diet when normally only roots are available. They can be processed by using conventional food storage methods. For example, try canning or freezing wild greens as you would spinach. To can, parboil the mixture until tender and then place into canning jars. Add cooking water, loosely seal, and process thirty minutes in a pressure cooker. Tighten lids and allow to cool. To freeze: parboil, drain, pat dry, and place in plastic bags.

DRYING WILD EDIBLES

Wild edibles can be dried and stored for later use in soups and sauces and as thickeners. To dry, tie the whole plant with a string and suspend it from a ceiling or in a warm ventilated area such as an attic, away from sunlight. Some may be spread on the floor to dry, but these plants may need to be turned over occasionally. Flowers and their seeds can be dried on screens but should not be crowded. When the plants are dried, crush their leaves and flowers into a powder and store it in air-tight jars. Some vitamins but few minerals are lost in drying. Plants used for medicinal purposes should be stored whole and not crushed until used.

Powdered dried edibles such as chickweed can be made into pasta. If you mix the dried powdered edible with flour and run it through a pasta maker, you'll produce a nutritious, uniquely-flavored green pasta similar to the artichoke pasta you see in the grocery store.

I enjoy drying my greens and using them in baked goods, pasta etc. all year for additional nutrition.

For additional information, read the chapter on food drying in my book, *Becoming Self-Reliant–How to Be Less Dependent on Society and the Government.*

FLOURS

To make bread products from seed flour, blend several kinds. For example, experiment by using acorn flour with wheat flour in equal parts or add some seed flour as a supplement to your usual baked products. Adding fruits may improve taste and moistness.

Flour can be made from a variety of wild edible seeds and leaves. For example, processed acorns, amaranth leaves, cattail roots, clover leaves and flowers/seeds, dandelion leaves, dock seeds and roots, grass leaves, kudzu roots, lamb's-quarters leaves, pine pollen, plantain leaves and seeds, prickly pear seeds, strawberry leaves, thistle roots, inner bark of trees, and purslane seeds–all can be ground into nutritious flour.

SPROUTING AND INDOOR GROWING OF WILD EDIBLES

If the seeds you find are more than you need for flour, consider seed sprouting as another source of nutrients. Sprouting can provide a continuous supply of greens and nutrients. During the winter, collect ripened seeds of shepherd's purse, sorrel, amaranth, burdock, and peppergrass. Allow the seeds to thoroughly dry and then store them in a dry place.

To sprout dried seeds, place in a fruit jar, cover with water, and soak overnight. Drain through a screen or cloth placed over the container. Refill jar with water to rinse and pour it off. Store the container in a dark, warm place. Repeat the rinsing daily until the seeds sprout. Place the sprouts on a sunny shelf until they develop a green color which increases the vitamin A and chlorophyll. Fresh sprouts in the diet add a boost of enzymes, vitamin C, B, and E. Several sources state that sprouting multiplies the seed's nutrients by 300 percent. The minimum amount of vitamin C needed to prevent scurvy can be supplied by 1 1/4 ounces of wheat or bean sprouts that are allowed to sprout 1 1/2 times the length of the seed. Sprouting in the dark seems to increase the B-vitamins while exposing to light during the last day increases the vitamins A and C. The liquid used for sprouting is nutritious although the taste may be objectionable.

An easy version of indoor gardening is to plant seeds in a flower pot containing soil similar to that where the wild edible was found. Place the pot near a sunny window, and collect greens as they grow! Chickweed and purslane work well in such an indoor planter.

Several health-conscious foragers I know grow rye this way and regularly clip and enjoy the new growth of fresh greens. One of the families uses the rye greens as a major portion of their diet, and they are robustly healthy. K.L.

Seeds of wild edibles provide an extension to the seasonal use of a fresh plant in the form of flour, cereal, and sprouts. Be aware that many seeds look large until the chaff is removed only to reveal a mere speck of a seed. Another tip: When you check out vacant lots and graded areas that have not been mowed, you may find domestic grain seeds. Look here for fescue, rye, and wheat from landscape seeding or from the wheat straw mulch.

CHAPTER II

IDENTIFYING WILD EDIBLES

Before we begin to identify plants, let me say that mushroom collecting is an area of expertise not covered in this text since mushrooms are low in nutritional value and I do not think wild ones are worth the risk of poisoning that might occur from erroneous identification. K.L.

Everyone deserves the opportunity of field training with an experienced forager before eating any wild plants he collects! Without this advantage, the novice should take first "finds" to a botanist or other expert for positive identification. Even after taking the precautions to get positive identification, the reader should initially taste only small amounts of the plant in question, then try to build up intake, gradually allowing the body to adjust to the new food.

In reality, it is difficult to find someone who is knowledgeable in foraging. The local extension agent may know of a herb or foraging club. Universities have classes in botany, and their professors may be willing to share their experience, or contact can be made with botanical gardens for help.

I was fortunate when I started, for I met several people who enjoyed foraging for wild edibles (one was a college biology professor). They had their own established "digs" and could, at the drop of a hat or "species," give exact directions to concentrated areas of specific edibles. What a warehouse of knowledge they were! I quickly became hooked on wild edibles as a renewable food source. K.L.

Outdoor Tip: One should not use an *unidentified* plant unless nutrition needs have become a matter of life and death. If ever this situation occurs, various military manuals recommend the following general guidelines for testing the target plant:

Beware of any plant that shows red in any part of its growth, in its fruits or stalk. Pokeweed acquires redness as the plant matures and becomes poisonous. Strawberries are normally safe, but some people become sick if they eat them. Rhubarb has a red stalk and the leaves can be deadly due to a concentration of oxalic acid. The tomato plant is in the same family as the deadly nightshade, and its leaves are dangerous, but the fruit, we know, is delicious.

Rules to follow:

Beware of trees and vines which do not have a clear sap when cut. A colored or white sap means danger.

Inspect it–Make certain the plant is not slimy.

Smell it–Crush a small portion. If it smells like bitter almonds or peaches, do not try it.

Rub it on–Rub the plant or its juice on a small, tender skin area. Discard it if a rash or swelling is experienced.

Cook it–Cook all plants if there is a question as to their edibility. Then take the following steps, stopping if there is any unpleasant reaction. First, take a very small amount of a cooked sample and place it on the lips. Wait five seconds, then chew a small amount and hold it in the mouth for five minutes. If it still has a pleasant taste, proceed to eat a small quantity. If the taste is really disagreeable, has a taste of almond, bitterness, or extreme acid, do not continue to eat it. However, you should recall that olives are bitter and grapefruit is sour, so an unpleasant taste does not automatically mean poison. Yet a burning, nauseating sensation, or very bitter taste is a warning of danger.

Finally, wait five hours without drinking or eating anything else during the time. A very small quantity of a poisonous food–**with the exception of mushrooms**–is not likely to prove fatal whereas a larger quantity may. Never try an unknown mushroom.

Hopefully, you will never be in a survival situation and need to try an *unidentified* plant. The mentioned "tests" will provide a reasonable safety level, but in the case of poison hemlock, extremely common along streams, you could die after the "testing" process. The key is not to rely on "rules of thumb", but to learn how to identify the specific plant and in some cases specific plant families.

I have never needed to eat an unidentified plant, but I have used these techniques when I have identified a "new" wild edible. By having a specific testing process to follow, I feel assured the appropriate precautions have been taken.

I once heard a saying about mushroom hunters: There are old mushroom hunters and bold mushroom hunters, but no old, bold mushroom hunters. K.L.

SEASONS

During different times of the year, wild edible plants go through various stages of edibility and availability. Normally, the early young growth in the spring is the most tender and tasty. As a wild edible matures, especially after blooming, it often becomes bitter.

By plotting the wild edible's growth patterns and habitats, one will create personal "digs" according to the plant's seasonal availability. Returning to the same area to observe a plant's various growth stages, one finds that it offers different foods throughout the year. For example, some can offer greens in the spring, seeds in the summer, and roots in the winter. This knowledge provided the forager a protective pride similar to watching your home garden grow.

LOCATIONS

Oddly enough, many beginners believe they need to head for the deep virgin woods to find wild edibles. In reality, however, one will find wild edibles in nearby disturbed open areas such as fields, and neglected farmland, gardens, and roadside areas. Of course, wooded areas are best for nuts and tree-based edibles.

Wild edible locations will vary. For example, while on assignment during the "Great Flood Of '93," flying out of Midway Airport in Chicago, I was curious if there were any wild edibles in the vicinity of such a large city. I found a vacant lot which held more wild edibles than I ever found before in one spot. Here grew chicory, curly dock, lamb's-quarters, dandelion, sow thistle, and burdock. What an opportunity for a person knowledgeable in foraging! K.L.

Considering locations, there are some warnings to be heeded. Do not pick wild edibles within 50 feet of well-traveled roads or where there may be any type of pollution. For some plants such as brambles and plantain, allow a one mile barrier. Plants beside a busy road may contain up to 200 times their natural level of pollutants. Be wary of plants growing near or in polluted water. If one must use such plants in a raw state, the bacteria–though not the pollution–can be killed with a mild chlorine bleach solution (use 1/2 teaspoon bleach per gallon of water and soak the plant for a half hour; then rinse it).

One will learn that areas disturbed by man are the best foraging areas. My first discovery was beside a back road. This was where I found my "special spot" of pokeweed. I had heard about pokeweed growing everywhere, but now I was having trouble finding it. Then one day along the roadway, I saw two women coming down a dirt bank out of a pasture, carrying several bags of large, leafy greens. Hoping it was pokeweed, I retraced their steps and identified these marvelous greens up close for the first time. Now that I recognize it, I see pokeweed in most every vacant lot that I pass. The second plant I discovered was chickweed. I was surprised to find that I had been digging it out of my flower bed for years as a weed! Then, at the edge of my garden I recognized a healthy crop of bittercress. Next, I discovered dock, and soon I was finding sorrel, wild carrots, and several varieties of wild lettuce. Having learned how to recognize wild edibles, one discovers them everywhere and in a variety of soils.

Of course, one will find some plants quite prolific while others will have spotty populations. For example, a country road near my home is thick with wild carrots though they are difficult to find elsewhere. In a similar way, at an intersection near my home there is a large stand of peppergrass, but I have not found it elsewhere. There are wild foods out there even if one cannot initially find them. Many times I have gone to other areas to find new plants, only to realize later that they were in my own backyard all along. K.L.

PRESSING AND DRYING PLANTS FOR FUTURE IDENTIFICATION

A dried plant is very convenient as an exact specimen to study when a second opinion is needed. An added benefit is the dried plant's retention of color and shape which can be helpful in a future season's foraging.

To press a plant, take the specimen and place it between two sheets of smooth paper. Place paper and specimen under a stack of heavy books. After two weeks, when the plant should be dried, apply several drops of glue or rubber cement along the stem and under the fragile leaf sections. Place the leaf on a sheet of paper and allow the glue to dry. Make notes on the margins of the mounting paper such as the time of year, the location of the find, and the type of soil there. Then, place the paper in a notebook for future reference. Photo or page protectors can be used, but they are not mandatory. You will have an actual-size sample of the plant which can be examined whenever there are any questions about details of the discovery.

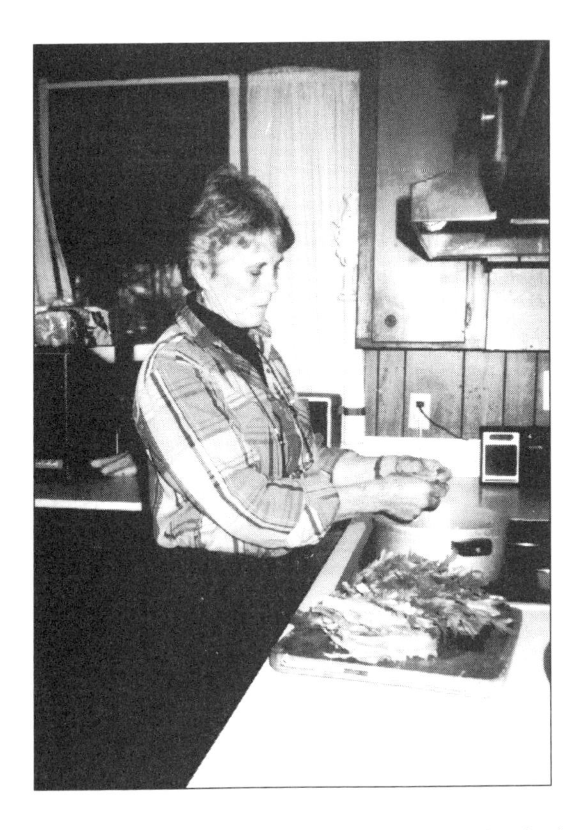

Wild edibles are being prepared to supplement other foods.

CHAPTER III

WILD EDIBLE NUTRITION

BETTER NUTRITION

Finding pleasure in identifying wild edibles is one thing; using wild edibles to benefit from their nutritional value is another! Wild plants contain essential vitamins and minerals and are rich in protein and carbohydrates. Some contain fat for calories and all provide roughage essential to keep the body in good working order. Unfortunately, most of the vegetables we see at the grocery store are grown in nutrient depleted soils where trace minerals cannot be replaced with commercial fertilizers. Also, these foods are not developed in the best interest of the end user. They are developed for commercial reasons such as appearance, shelf life, shipping, and ability to be mechanically harvested. Wild foods can have 400 percent more vitamins and an untold amount of trace minerals as do organically grown foods.

A proper diet uses a wide range of elements to provide the correct amount of nutrients and calories to maintain life. These must include fats, protein, carbohydrates, minerals, trace elements, and vitamins. All these essential elements may be found in wild edibles. Understanding a plant's nutritional benefits is important, particularly when food or funds are in limited supply.

CALORIES

We will begin our study with calories and their function. The U.S. Department of Health and Human Resources suggests two thousand calories per day as an average requirement. Following this guideline is very important in a food shortage situation where the lack of calories can cause sickness and death. The details of various foods' caloric level follows. Wild edibles are listed in bold print.

CALORIC VALUES

SOME COMMON AND SOME WILD FOODS	Calories per 100 mg. or 3 1/2 oz.
Blackberries, fruit, raw	**52**
Milk	61
Potatoes, baked with skin	109
Rice, brown, cooked	112
Kudzu roots	**113**
Soybeans	141
Eggs	155
Beef, chuck, blade, 1/4 fat trim	237
Chicken, white, roasted	239
Maple, syrup	**262**
Bread, white	361
Oaks, acorn, raw	**369**
Arrowhead, roots	**413**
Peanuts, with skin	567
Walnuts, dry	**607**
Pecans, dry	667

Source–See Nutrition Analysis Chart - Appendix C

Outdoor Tip: Do not rely on the easiest source of food. A balanced diet of nutrients and calories is very important. There are documented cases of survival-situation hunters starving to death on a regular diet of rabbit due to the rabbit's leanness.

One must remember that caloric needs double in strenuous work, and that anxiety as well as physical effort uses calories. Calories are not produced equally by all kinds of foods. If foods are limited, one should seek those that have the most calories. For reference, one gram of fat produces double the calories of the same amount of protein or carbohydrate:

Carbohydrates, 1 gram	–	4 calories
Protein, 1 gram	–	4 calories
Fat, 1 gram	–	9 calories

Although fats, proteins, and carbohydrates make up an important part of the body's dietary needs, an adequate vitamin and mineral intake must also be ingested for balanced nutrition. Many people are quite surprised to see the comparison of wild edibles with domestic plants as providers of these elements.

CARBOHYDRATES

Carbohydrates serve to provide energy for muscular contraction and other body functions. They are easily converted into energy by the body and do not require a large water intake. Carbohydrates are the most rapidly utilized sources of energy and normally supply two-thirds of that used by the body. In a diet they reduce protein metabolism so that a large amount of meat is not necessary. Carbohydrates do not contain vitamin B. Excess carbohydrates may cause constipation in an unbalanced diet.

There are two types of carbohydrates:

Sugars–found in sugar, syrup, honey, and fruits.

Starches–found in cereals, roots, and tubers. One should cook roots and tubers to break down their carbohydrates for bodily use.

PROTEINS

Proteins are the basic building blocks (amino acids) for construction and regeneration of the body tissues. In other words, one takes in protein foods to extract from them the amino acids to build other specific proteins needed for body growth and renewal.

Even if all other amino acids are present, the absence of one component will greatly limit the protein available for use by the body. Proteins have an advantage of being available from both vegetable and animal sources. They are found in meat, fish, eggs, dairy products, plants, nuts, and grains. Animal protein contains all the amino acids we need while plant foods do not, unless a wide range is eaten.

Proteins are divided into two basic groups:

Complete Proteins–These are foods that have a balance of amino acids. For example: meats, milk, poultry, eggs, cheese, and fish.

Incomplete Proteins–Grains, nuts, seeds, and legumes.

A number of dishes effectively combine proteins to maximize utilization. They may use rice or wheat with beans, for instance. A good guideline is to mix by weight 2 parts grain and 1 part legume.

The body also conserves protein to some extent. If it lacks sufficient carbohydrates, some proteins can be converted into dextrose to yield energy, but the amount of such conversion is limited.

FATS

When fats are oxidized by the body to liberate heat, they supply energy. Stored body fat is an immediately available source of energy. Unfortunately, however, one cannot be supported on fats alone except for vitamin E. They do not supply other nutritive substances such as vitamins which might be found in food containing protein or carbohydrates. The important point is that a small amount of fat is necessary; otherwise, a deficiency can lead to severe disease symptoms including renal (kidney) lesions. Weight-conscious people must keep this fact in mind. In a food shortage situation, even commercial vegetable oil would add 100-120 calories per tablespoon to the diet.

Fats are a concentrated source of energy. Before they can be absorbed by the body, they require a lengthy digestive process. This digestive process requires an adequate intake of water; therefore, if water is scarce, fat intake should be controlled. Fats are found in animals, fish, eggs, milk, nuts, and some vegetables.

VITAMINS & MINERALS

Because the body cannot manufacture vitamins, essential vitamins must be supplied in the diet. Vitamins serve as essential catalysts for metabolic reactions in the body. Though required in only very small amounts, vitamins, especially vitamin C, must be included daily in one's diet. Some vitamins are water soluble so they are lost when prepared with water.

NUTRIENT LOSS FROM COOKING DANDELION

Nutrients	Loss %
Calories	-27
Protein	-26
Fats	-14
Carbohydrates	-30
Calcium	-25
Phosphorus	-36
Iron	-42
Sodium	-42
Potassium	-42
Vitamin A	-16
Thiamin	-32
Riboflavin	-33
Vitamin C	-49

Source: Calculated from Composition of Foods USDA Vegetables 8-11 1984

It makes sense that we should make every effort to utilize the whole plant, to use a minimal amount of water, and to consume the cooking liquid.

Wild edibles often contain more vitamins and minerals than commercially marketed plants and products, as you will see from the charts that follow. A carrot is well noted for its vitamin A content, but wild dandelion and lamb's-quarters are also rich in vitamin A. Orange juice is regularly used for vitamin C, but wild violet, lamb's-quarters, dock, and amaranth are also good C sources. For calcium needs, lamb's-quarters and shepherd's purse are excellent. Furthermore, wild edibles are plentiful, free, and, like those vegetables in your own garden that are fresh and nutritious, they are just outside your backdoor!

Minerals are important for proper nutrition because they contribute to the supporting framework of the body, enter into the formation of organic compounds, and exert an influence upon the functions of tissues. Due to food preparation methods, mineral elements naturally present in foods are often lost. Two common practices which negatively influence mineral content:

1. Discarding fruit peels and hulls of grains which results in a loss of iron, calcium, and phosphorous.

2. Throwing away the cooking liquid discards much of the foods' mineral and vitamin content of foods which goes into solution during cooking.

RECOMMENDED DIETARY ALLOWANCES: VITAMINS

Before we look at the vitamin offerings of specific plants, let us review the recommended dietary allowances for important vitamins such as vitamin A, Thiamin, Riboflavin, Niacin, and vitamin C.

RECOMMENDED DIETARY ALLOWANCES (RDA)
Vitamins

	Vit A (RE)	Thiamin (mg)	Riboflavin (mg)	Niacin (mg)	Vit C (mg)
Male	**1,000**	**1.5**	**1.7**	**19**	**60**
Female	**800**	**1.1**	**1.3**	**15**	**60**

Source: Recommended Dietary Allowances. The National Academy of Sciences, Washington, D.C. 1989

VITAMIN A
Recommended Daily Allowances (RDA)
Male 25-50 3,300 IU (International Units)

FOODS (100 grams)	Mg. per
Carrots	28,129
Dandelion	14,000
Lamb's-quarters	11,600
Dock	4,000

Source: Composition of Foods USDA Vegetables 8-11 1984

The following charts list the commonly accepted sources of vitamins followed by a wild edible alternative. The vitamins that may be deficient in a food shortage situation are A,C, and D.

VITAMIN C
Recommended Daily Allowances (RDA)
Male 25-50 60 mg.

FOODS (100 grams)	Mg. per
Orange Juice	53
Violet Leaves	210
Lamb's-quarters	80
Dock	48
Amaranth	43

Source: Composition of Foods USDA Vegetables 8-11 1984

PROTEINS AND MINERALS

The following charts review the recommended dietary allowances for protein and minerals. To help in relating wild edibles to domestic foods, the following comparisons will list the commonly accepted sources of proteins and minerals followed by the wild edible alternative.

RECOMMENDED DIETARY ALLOWANCES (RDA)
Protein & Minerals

		Protein (g)	Calcium (mg)	Phosp. (mg)	Iron (mg)
Male	(25-50)	63	800	800	10
Female	(25-50)	70	800	800	18

Source: Recommended Dietary Allowances. The National Academy of Sciences, Washington, D.C. 1989

Frequently, wild edibles would need to be supplemented with meat or bread to provide adequate protein. However, minerals may be completely supplied by wild edible alternatives.

PROTEIN
Recommended Daily Allowances (RDA)
Male 25-50 63 g.

FOODS: (100 grams)	Mg. per
Beef, chuck, blade, 1/4 fat trim	31.0
Walnuts, dried	24.3
Oak, acorns, raw	6.2

Source: Composition of Food USDA Vegetables 8-11 1984

IRON
Recommended Daily allowance (RDA)
Male 25-50 10 mg.

FOODS: (100 grams)	Mg. per
Beef Liver (cooked)	6.3
Sorrel, Sheep	5.0
Shepherd's Purse	4.8
Dandelion	3.1
Dock	2.4
Amaranth	2.3

Source: Composition of Foods USDA Vegetables 8-11 1984

CALCIUM
Recommended Daily Allowances (RDA)
Male 25-50 800 mg.

FOODS: (100 grams)	Mg. per
Milk	119
Lamb's-quarters	309
Amaranth	215
Shepherd's Purse	208
Dandelion	187

Source: Composition of Foods USDA Vegetables 8-11 1984

VITAMIN & MINERAL SUPPLEMENT TIP: Try making your own vitamin supplements. Begin by filling empty gelatin capsules with a dry mix of crushed and powdered wild edibles. You may have to shop around to find a store that carries the capsules. Smaller drug stores carry capsule sizes from #00 (very large) to #4 (very small). I would suggest the #00 size for best results. Begin by taking one capsule at each meal.

Cattail roots and shoots provide nutrition and carbohydrates.

The study of nutrition obviously can be rewarding. Accordingly, you may want to visit your local library and learn more about "vegetarian" techniques and how to balance protein, carbohydrates, and fats. As you begin to develop complimentary combinations of foods to best use the amino acids, you will reduce nutrient waste and prevent a dependence on perishable foods such as meat and milk. Remember, a proper diet includes calories, fats, protein, carbohydrates, minerals, trace elements, and vitamins. All these elements may be found in wild edibles.

Understanding a plant's nutritional benefits is important particularly when food, or funds, are in limited supply. See Appendix C for Nutrition Analysis Chart.

During stressful times, plan to take vitamin C and multiple vitamin and mineral supplements to insure adequate intake.

MAXIMIZE NUTRIENTS

One should try to maximize the available supply of nutrients.The stems of plants produce fewer nutrients than the roots, shoots, and leaves, but the stems may generally be used if necessary. They would be discarded normally when plants are in abundance.

If the wild edible has become woody or fibrous, it can be blended and pressed through a sieve. The resulting liquid goes easily into a soup or a cream sauce. Mature leaves, when finely chopped and parboiled, can be added to stews.

Outdoor Tip: The raw plant can be chewed, the liquid swallowed, and its pulp spit out.

Cooking liquid (exception: that from boiling pokeweed) should be used to reclaim nutrients leached into it. Bitter liquid can be mixed in with something else. In cold climates, the liquid can be frozen for use later in soup.

ROOTS AND TUBERS

Roots and tubers are invaluable nutritional food–they have good nutritional value plus starch and calories. Roots are at their starchiest between autumn and

A large metropolitan grocery store offers wheat grass for sale.

spring. The vitamins of some tubers occur near the surface of their skin, so it should not be peeled away unnecessarily. Most roots and tubers are safe but must be cooked to be digestible. (Even the common potato has little nutritional value until cooked.) To reduce nutrient loss and cooking time, some of the roots may be cubed first.

Many plants and vines have tubers. This general principle can help when you are foraging: If the leaf's veins radiate from the joint of leaf and stem rather than from the main center vein, dig the plant for it is a likely source of tubers.

INNER BARK, BUDS, AND YOUNG SHOOTS

The inner bark, buds, and young shoots of many trees are edible also, for example: basswood buds, pine, birch, maple, spruce, fir, aspen/poplar (parboiled in several changes of water to remove bitterness,) willow, and cottonwood. (Some native American names mean "tree eaters." This fact suggests another bit of survival knowledge our "civilized" culture has largely ignored.)

SEEDS

Survival Tip: In a survival situation, even very small and difficult-to-separate seeds, can be eaten if the chaff and seed are ground together and mixed (if no yeast is available,) with water to make bread or pancakes. Slowly baked, the result will be somewhat heavy but palatable and nutritious.

BALANCED DIET

A balanced diet is very important. As stated before, lean meat alone, for instance, does not provide all the necessary nutrients.

Survival Tip: It is important to keep calm and relaxed when food is scarce, and to take in adequate calories (fats) as well as other nutritional elements.

CHAPTER IV

GUIDES TO WILD EDIBLES

For obvious reasons, the beginning forager should consider concentrating on specific edibles that offer the greatest opportunity for success. This chapter emphasizes plants that are not difficult to identify and which are widely found in most of the United States, Britain, and Northern Europe. Typically, these plants grow in concentrated areas, thus are relatively easy for the beginner to find. I would suggest these "beginner" plants: Dandelion, dock, plantain, poke and wild lettuce.

Before eating any wild plants, a collector should have developed expertise through field training with an experienced forager. If this is not possible, he should take the "finds" to a botanist or other expert for positive identification. Even after taking these precautions, the novice should initially taste only small amounts of the plant in question, then try to build up intake, gradually allowing the body to adjust to the new food.

Foraging can become a whole world of fascinating and practical involvement. It does, of course, as with any skill, require hands-on experience. When I first became excited about wild edibles, I looked for them everywhere with minimal finds. We deserve rewards for our efforts, so I advise new foragers to begin searching for just one or two wild edibles that are easily identified and known to be available in the area.

During your experimentation, you will find that you prefer certain edibles to others. In addition, you will find notably different wild edible concentrations in different parts of the country. Practice now while you can leisurely develop your own "digs", preparation techniques, and even transport favorites to your garden.

One bit of warning is important here: Avid foraging fans aren't the safest drivers! One of my biggest problems is balancing driving safety with watching for roadside plants. Keep things in perspective as you enjoy a new adventure and a beneficial hobby. K.L.

WILD EDIBLES

This section will review the specific details of selected wild edibles. The plants identified are listed in alphabetical order. The reference to range of location is pertaining to the United States, Britain, and Northern Europe unless noted.

South Florida does not have an abundance of wild foods as do the other states, but I noticed several plants while there with Hurricane Andrew. In Miami, I found a nice patch of broadleaf plantain under the rail system near 67th street NW and 27th Avenue and large oak trees (for acorns) downtown. While crossing the Everglades on Highway 41, I saw maple trees, and in Miami Springs I found purslane growing in the sidewalk. Then, there was amaranth in Miami Beach under an overpass on Indian Way, peppergrass in Naples, and wood sorrel in Key West. These plants may have sparse populations, but you can find them. Good hunting! K.L.

In Britain and Northern Europe, the climate is perfect for wild edibles. Near London I found wild carrot, dandelions, and Canada thistle. In Hyde park was chickweed and broadleaf plantain. West London had curly dock, narrow-leaf plantain, and sow thistle. Salisbury had bull thistle while Chester had wood sorrel and clover. Later in Moffet, Scotland, I found wild rose and prickly lettuce. In Abbotsford, I found sheep sorrel and bittercress. On my way back to London from Scotland, I saw burdock, wild mustard, and lamb's quarters. So, if you have an opportunity to make the trip, take your camera and this book! K.L.

Amaranth

IDENTIFICATION

Amaranth is an annual plant with many variations. Amaranth is stout in shape and has rough, hairy stems. Some species can have small thorns where the stems meet the main stalk. Generally amaranth is freely branching with noticeable veins in stems and can grow to six feet in height. Amaranth is fast growing, drought resistant, and abundant in a variety of soils. Some call this plant pigweed because it likes the rich soils around pigpens, and many farmers use it for pig feed. The plant is green, but some varieties have a red or purple tinge on branches and flower petals. Leaves are alternate (rather than opposite) on the stem, grow to five inches long, and are smooth to rough and sometimes hairy to the touch. The leaves can be oval to lance-shaped and wavy to saw-toothed, depending on the species. The midribs and veins are very noticeable from the bottom of the leaves while the lower leaves are purple on the underside. Flowers are small and greenish but cannot be described as blossoms. Abundant small, black shiny seeds are produced at maturity. Roots are red. Because of growing interest in amaranth, many garden seed catalogs carry a domestic variety.

Several years ago after reading about amaranth grain in Organic Gardening magazine one winter, I began to look for it in my travels. Every farmer I talked to knew it by the name "pigweed." Unfortunately, there was none growing for identification during that time of the year. Due to a heavy travel schedule, I did not plant my garden the next spring. During mid-summer I saw that it was overgrown with a large, tough plant. I had found amaranth in my own garden after looking fruitlessly elsewhere! K.L.

RANGE AND ENVIRONMENT:

Amaranth is widely found in rich soils around barns. Also, it is found in waste ground, along highways, and in other disturbed soils.

Warning: Amaranth contains oxalic acid as does spinach. According to some researchers, amaranth can inhibit the absorption of calcium in the body if eaten in large quantities. Newer research may be negating this opinion; for now, consider a normal serving to be safe.

Amaranth

Amaranthus spp.

IDENTIFICATION BRIEFS

Species:	*Amaranthus spp.*
Size:	Up to six feet
Other Names:	Pigweed, Red-Root Pigweed, and Careless Weed
Blooms:	Green
Sun Required:	Full sun
Flower Color:	Green
Type:	Annual
Propagation:	Seed

LEAVES:

Many feel that amaranth tastes better than garden spinach.The leaves are gathered during the spring and eaten raw or cooked like spinach. Mature leaves become tough and bitter. Use young, tender amaranth leaves to moderate stronger flavor of greens such as dandelions and wild mustard. Leaves can be dried and ground into a flour and mixed with other flour.

Outdoor Tip: Because of its ability to produce a soapy lather, the leaf is used for washing clothes.

SEEDS:

The seeds are gathered during late summer and fall. They are high in protein and, when mixed with corn, make a complete protein. The small seeds are parched and ground into flour or mixed in with soups. The flour can have an “off” flavor which is reduced if seeds are parched before grinding or when ground seeds are mixed with other flour. You will find that amaranth flour is used in several brands of bread and breakfast cereals found in health food stores.

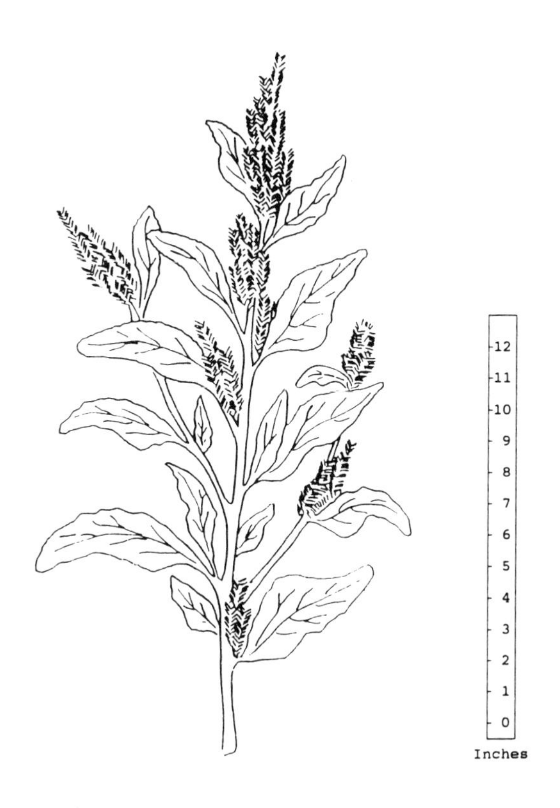

Amaranth

Arrowhead

Sagittaria spp.

Arrowhead

IDENTIFICATION:

Broadleaf arrowhead has distinguishing broad, arrow-shaped leaves while grass-leafed arrowhead is lance-shaped. Both have toothless leaves and can grow to three feet tall.

It usually grows in clumps, and the roots have small, edible potato-like tubers. Flowers are white with three petals in groups of three. The plant dies back in the winter and is hard to find unless its location has been marked during the prior season.

RANGE AND ENVIRONMENT:

Arrowheads are found everywhere; broadleaf arrowhead is more widespread, while grass-leafed arrowhead is usually found in the Eastern part of the country. Both prefer shallow water and muddy areas.

IDENTIFICATION BRIEFS:

Species:	*Sagittaria spp.*
Type:	Perennial
Other Names:	Wapatoo
Size:	Up to three feet
Flower Color:	White
Blooms:	Summer
Sun Required:	Full sun to semi-shade
Propagation:	Division

TUBERS:

The tubers are collected fall through winter. They are a good source of carbohydrates and were used by Native Americans as a primary vegetable. The tubers grow in the mud at the end of long underwater roots

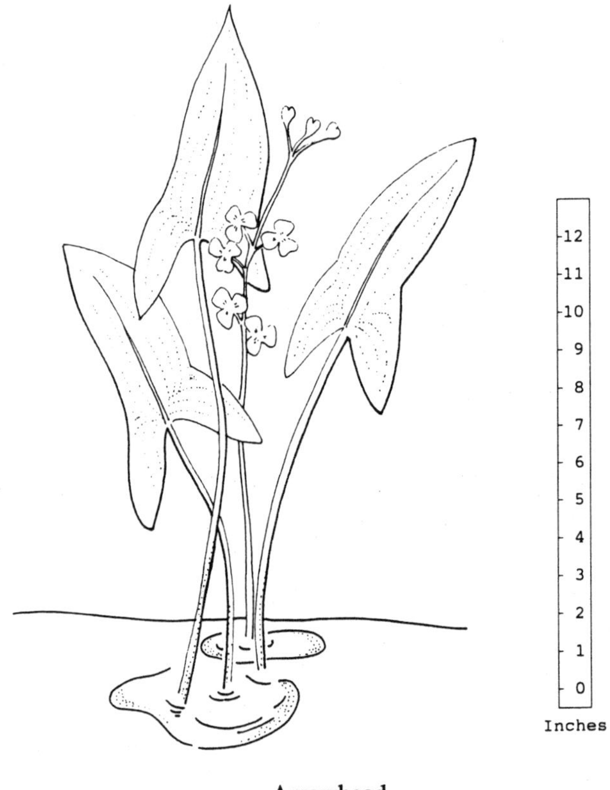

Arrowhead

sometimes remote from the main plant. In shallow water, one may find tubers by tracing the roots by hand. The tubers vary in size and can grow as large as a "new" potato. To harvest, loosen the tubers using a hoe or bare feet, and they will float to the surface. In mucky soil with an intermittent water source, the roots and tubers can become very matted and difficult to impossible to find.

Arrowhead tubers are edible raw after peeling, but they do not have a desirable taste. To improve the taste, cook them like potatoes with the peel. Cooking also protects against bacteria from the source water and makes them digestible. After cooking, remove the peel and add the food to a salad or mash and fry like hash browns.

To store, arrowhead tubers can be strung on a string and dried for later use.

Bittercress

IDENTIFICATION:

Pennsylvania bittercress has a large rounded terminal lobe (tip of leaf) and smaller round opposite leaves. The leaves are equally spaced as they approach the base of the stem. They are dark green, somewhat shiny, and 1-3 inches long. When bittercress begins to flower, the flower stem will reach six inches to one foot in height. The white flowers have four petals and are very small and clustered at the tips of the stems. When mature, the flowers are followed by slender seed pods about one inch long. Bittercress is very similar to watercress.

For many years in the South, I heard about a plant called creesy. Then one day I asked a coworker

Bittercress

Cardamine pensylvanica

who lived in the north Georgia mountains about them. He said he cultivated them in his garden and would bring me some. When I received several plants, I set them out in my garden for observation. While returning to my house, I noticed a new weed in the flower beds. Sure enough, it was creesy–or more properly called–bittercress. Now we have bittercress greens available whenever I clean out my flower bed! K.L.

RANGE AND ENVIRONMENT FOUND:

Bittercress is found widely. Usually bittercress prefers to be near moist meadows or streams. It can become a "lawn weed" in yards or parks where there is some protective shade and the soil is not acidic.

IDENTIFICATION BRIEFS:

Species:	*Cardamine spp.*
Other Names:	Creesy and Creases
Plant Size:	1-3 inches
Flower Color:	White
Blooms:	Fall to early spring
Sun Required:	Full sun to semi-shade
Propagation:	Seed
Plant Type:	Annual

LEAVES:

The leaves are found late fall to early spring. They are used fresh in salads or cooked like domestic greens. If the pungent taste is not desirable, parboil the green for several minutes, drain the water, add fresh water, and cook to desired tenderness. Bittercress is a good substitute for watercress or may be mixed with milder edibles to add zest.

To freeze, blanch for two minutes, drain, cool, pat dry, and place in plastic bags.

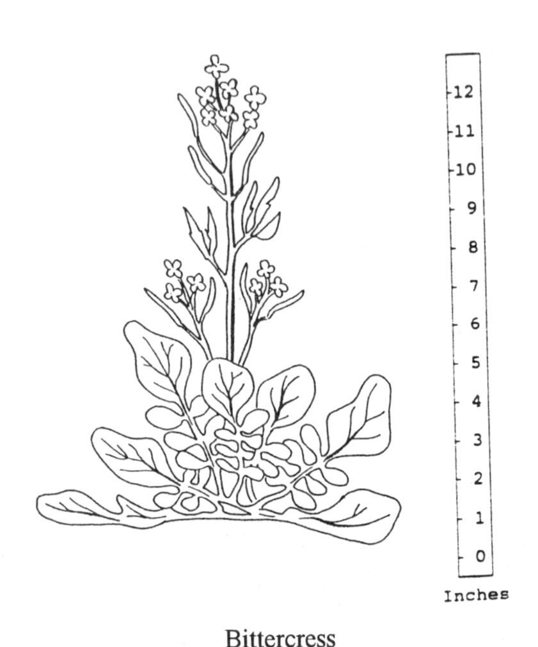

Bittercress

Brambles

IDENTIFICATION:

Brambles are well known by everyone who has had clothes caught or limbs scratched by them. They grow erect to five feet tall and are sometimes trailing in nature. Many species in this genus, range from black-

Brambles

Rubus spp.

berries to raspberries, and they are all edible. The flowers, white to pink, are followed by a juicy fruit. Leaves are compound and saw-toothed. We will discuss blackberries, dewberries, and raspberries.

Blackberry–Blackberry stems are angular, erect, and armed with thorns.

Dewberry–Dewberry stems are more vine-like and low-growing, and the fruit ripens earlier than blackberries.

Raspberry–Raspberry stems grow in an arch, are usually powdered-looking and round. The berries when picked leave their core attached to the stem.

RANGE AND ENVIRONMENT:

Brambles are widely found in sunny places. They prefer roadsides or other disturbed ground, and are found along the edge of woods.

IDENTIFICATION BRIEFS:

Species:	*Rubus spp.*
Other Names:	Blackberry, Dewberry, Raspberry, and Running Berry
Plant Size:	Trailing to five feet tall
Flower Color:	White to pinkish
Blooms:	Spring
Sun Required:	Full sun
Propagation:	Seed, division
Plant Type:	Perennial

LEAVES:

Bramble leaves are used spring through summer. The dried leaves may be steeped and used as a tea. For a tea, a tablespoon of dried or a handful of green leaves should be adequate. The young leaves can be eaten raw or cooked, but are not very tasty. I find the leaves are hairy and are best chopped and cooked. Always avoid wilted or old leaves because they may make you sick.

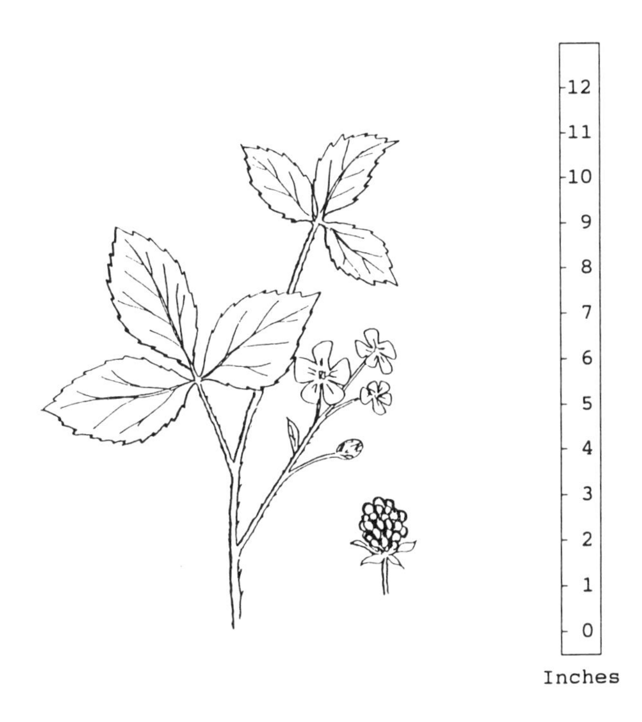

Brambles

SHOOTS:

The shoots are gathered during the spring. Use the fresh, young, bitter shoots alone or in salads. Try scraping before use and cook like asparagus to reduce bitterness.

BERRIES:

Bramble berries, collected during the summer, are popular fresh or when used to make juice, jam, or jelly.

Burdock

IDENTIFICATION:

Burdock is a biennial (flowering the second year) bushy plant with a stiff branching nature. It grows 3-5 feet tall with first-year leaves growing in rosettes (a ring of leaves radiating from the center.) Second-year leaves grow alternately on the stems. The leaves are large, oval, toothless, and rough with a woolly underside. The leaves have predominant veins and resemble an elephant ear.

The purple flower heads are surrounded with hooked burrs during the second year of growth. There are two types of burdock:

Great Burdock–Great burdock has large burrs to 1 1/2 inches in diameter with large oval leaves up to one foot in width. Lower leaves can be heart-shaped.

Burdock

Arctium spp.

Common Burdock–Common burdock has smaller leaves, burrs, and a shorter stalk than great burdock.

RANGE AND ENVIRONMENT:

Common burdock is found throughout the world, but is sparse in the Southeastern United States. Great burdock is found mostly in the Northeast United States, Britain, and Northern Europe. Both are found in disturbed soils such as along roadsides.

IDENTIFICATION BRIEFS:

Species:	*Arctium spp.*
Other Names:	Great Burdock, Common Burdock, and Beggar's Buttons
Plant Size:	3-5 feet tall
Flower Color:	Purple
Blooms:	Summer to fall
Sun Required:	Full sun
Propagation:	Seed
Plant Type:	Biennial

LEAVES:

The leaves are gathered in the spring. Cook young burdock leaves or use fresh in salads.

Though burdock is enjoyed by some, I find it fuzzy when raw and when cooked, undesirably chewy. K.L.

Outdoor Tip: The larger leaves of burdock can be used to wrap fish and meat for pit cooking and as temporary roof shingles for survival shelters.

ROOTS:

First-year burdock roots are collected during early summer to fall but are best young when not too bitter or tough. Second-year roots are very fibrous and stringy. The roots are peeled to remove the thick rind and parboiled for half an hour. Some like to add a pinch of soda to the first cooking water and then serve with butter. Drain, add new water for final cooking until tender, Or, the cooked roots can be cut into pieces and fried in butter. I slice the roots, brush lightly with olive oil, cover with aluminum foil and roast in the oven. Remove foil during the last few minutes for a crisper texture. In Japan

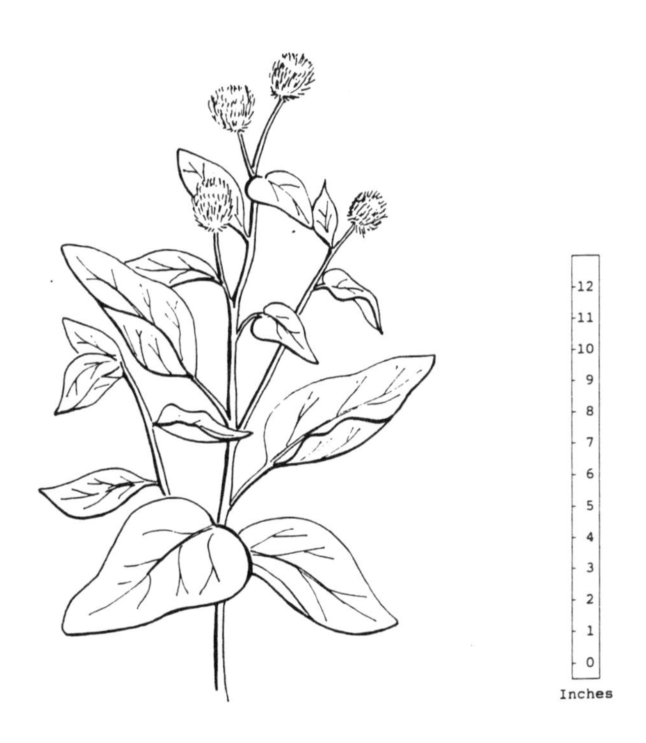

Burdock

it is called "gobo" and is considered a choice food. The extra roots can be dried and stored or ground up to be used later as soup thickener. Burdock roots are a good source of carbohydrates.

STALKS:

The stalks are found spring to summer. Before the flowers bloom, the young leaf and flour stem stalks are peeled to remove all bitter green skin and then used as asparagus. The stalks can be dried and stored or ground up and used later as a soup thickener.

SEEDS/BURRS:

The seeds are gathered during late summer. Cook them or bake with the young burrs in a casserole. Burdock seeds can be ground into flour.

Outdoor Tip: Burdock burrs were the prototype for the velcro zipper. They can be used to fasten together clothing, sleeping bags, etc.

Carrot, Wild

IDENTIFICATION:

Wild carrot is a biennial that grows 2-3 feet tall with a hairy stem. Leaves are deeply dissected (finely cut). It is easy to spot from the road because the white umbrella flower head stands out among the other plants. A 4-6 inch round, umbrella-shaped cluster of tiny (1/16

Wild Carrot

Daucus carota

inch) flowers make up the second-year flowerhead. The flower head known as Queen Anne's lace stands at the end of a slender stalk that is round and somewhat rough with fine hairs. Later, the dried flower's clusters create a circular pattern that resembles a bird nest.

Caution: Several identifying keys separate wild carrot from the toxic poison hemlock or fool's parsley. For example, wild carrot normally has one blue/purple flower in the center of the flower cluster; the root has a carrot smell; and it prefers a dry growing environment. Poison hemlock or fool's parsley is found in wet areas.

RANGE AND ENVIRONMENT:

Wild carrot is found widely in fields, waste ground, disturbed soils, and along highways.

IDENTIFICATION BRIEFS:

Species:	*Daucus carota*
Other Names:	Queen Anne's Lace and Bird's Nest
Plant Size:	2-3 feet tall
Flower Color:	White
Blooms:	Summer to fall
Sun Required:	Full sun
Propagation:	Seed
Plant Type:	Biennial

LEAVES:

The leaves are gathered during the spring. The tender young leaves can be cooked as greens or used fresh in salads. Also, they are good added to stews for seasoning. Some like to parboil them a few minutes, pour off the liquid, and then cook until tender. Leaves can be dried and later added to other foods as seasoning.

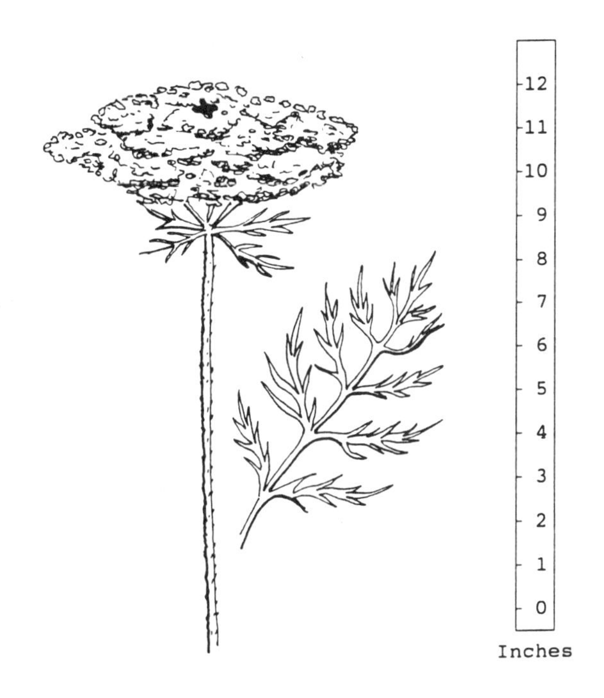

Wild Carrot

ROOTS:

The roots are collected during the spring. First-year root growth (those plants not blooming) are tender and are used like domestic carrots. When using older roots, remove the hard core after cooking, and allow the soft pulp and the juices to flavor the other foods cooked.

FLOWERS:

The flowers are used during the summer. The fresh flower heads can be battered and fried like fritters, or can be used to make jelly.

WILD CARROT JELLY

4 cups wild carrot flowers
1 teaspoon lemon extract
1 package pectin
5 cups sugar

Chop wild carrot flowers and place in a bowl. Pour one quart of boiling water over the flowers and set in refrigerator overnight. The next morning, strain and add one teaspoon of lemon extract and one package of pectin. Bring the mixture to a full boil. Add sugar, then bring to second boil, stirring constantly. Allow to boil for one minute. Skim, then pour into sterilized jars and seal. Same recipe can be used for many types of flowers, including roses, violets, kudzu, and dandelions.

SEEDS:

The small seeds are found during late summer. They can be easily gathered and used as a seasoning for soups and stews. In addition, the seeds can be used like caraway seeds to top breads.

Cattail

IDENTIFICATION:

Cattails have long firm stalks, each topped with a dense sausage-shaped seed cluster that appears after the flowers fall off. The seed stalk grows to eight feet in height.

RANGE AND ENVIRONMENT:

Cattails are found widely in wet ground, ponds, and swamps where they create large colonies.

IDENTIFICATION BRIEFS:

Species:	*Typha spp.*
Other Names:	Cossack Asparagus and Bulrush
Plant Size:	Up to eight feet tall
Flower Color:	Green turning to brown
Blooms:	Late spring
Sun Required:	Full sun
Propagation:	Division, seed
Plant Type:	Perennial

Survival Tips: Cattails are a year-round survival food abundant in ditches, through low water areas, and around ponds.

Cattail

Typha spp.

TUBERS/ROOTS:

Gather cattail roots any time of the year. They are best when gathered from late fall through early spring when the starch is concentrated in the roots. After spring, the roots slowly shrink, harden, and become ropelike. Studies show that the roots are very good at absorbing chemical pollutants so do not collect in such water.

Cattail

When baking, many prefer to mix cattail starch/flour with an equal mixture of wheat flour. To process the starch out of the roots, crush them in cold water. Next, pour the liquid through a sieve to separate out the fiber. Allow the liquid to sit until the white starch settles to the bottom, and then pour off the clear surface liquid. Add new water, stir, and repeat the process several times until all the fiber and particles are removed. After the final pouring off of the liquid, the remaining starch can be used wet as thickener or stored after being dried in the sun. Or roots can be dried in the sun for a week until they crack. Grind into flour and sift out the fibers.

I made my first cattail starch while on duty after Hurricane Hugo in Sumter, South Carolina. During the day I collected roots and at night in the privacy of my motel room, I processed the roots into starch. With a minimum of effort, I had a motel glass full of liquid starch. The most difficult part of the process was making sure no one saw me unload the roots into the motel room! Also, I sometimes wonder what the maid thought the next morning about the waste basket full of leaves and roots. K.L.

Outdoor Tip: Cattail roots are a rich source of carbohydrates. To prepare a gruel, scrape, clean, cut into pieces and remove large fibers from the cattail roots, then boil them to the desired thickness. To make a sweetener, crush roots in water, then strain and boil them down into a syrupy liquid of marginal sweetness. To make a flour, dry the roots thoroughly, skin them, remove fibers, and pound.

CORMS (Sprouts)

The corms are gathered during late summer to winter. The white horn-shaped corms grow along the root's length at the base of the cattail. Cook new corms like a potato. At the base of each sprout is a lump of tender starch material that is also cooked. The white baby shoots, emerging from the horizontal rhizome, are tender and tasty in salads.

SHOOTS:

Gather shoots during the spring. The green shoots grow out of last year's dead stubble and are easy to collect when about two feet tall. Reach down to the base of the leaves and pull while twisting the shoot. The top of the plant above the roots will break off leaving the green leaves and the white inner shoot. Peel off the outer layers until the tender white core is reached. Boil or steam for ten minutes if you like them crispy or boil longer to make them softer. They have a flavor similar to cucumbers. Also, shoots may be chopped into 1/2 inch lengths and added to a stew. To freeze, blanch for two minutes, drain, pat dry, and place in plastic bags.

I was once swapping wild foods information with a family that was knowledgeable in food storage techniques. They were teaching me food storage while I was showing them how to use cattail shoots. I had always boiled my shoots, but that night I decided to stir-fry them. I was sure I had enough for a meal, but right before our eyes, the meal dissolved into nothing. It would seem the shoots are mostly water, and the heat had evaporated the water. Another case of the importance of practicing what you "think" will work before going on stage with it! Now, I only boil my shoots. K.L.

Survival Tip: Water can be obtained from young cattail shoots by chewing them to remove the liquid and spitting out the pulp. They make an especially good water source when they are growing in moist ground with no surface water to contaminate them.

Caution: Learn to distinguish young cattail shoots from the poisonous look-alike iris shoots. To do so is very simple. Iris shoots have sharp-pointed leaves unlike the rounded tips of cattail leaves. Later, iris produces colorful flowers in contrast to the sausage-shaped seed cluster of a cattail. Iris roots are unpleasant tasting in contrast to the bland taste of cattail roots.

Survival Tip: The root fibers and cattail leaves can be twisted into cordage. Cattail leaves make an excellent weaving material for mats and baskets. Equally important, dried cattail flower stalks make good dry bedding which can be found most of the year.

FLOWER SPIKES:

Gather flower spikes during late spring. The immature flower spikes or flower buds should be gathered while still young and green. Husk papery bud sheath and boil in salty water for about ten minutes or until tender (some prefer to cook them still in the husk). Serve with butter and eat the flower head from its tough inner core like corn on the cob; otherwise, scrape or cut it off the core. The flavor is similar to corn.

SEEDS:

Cattail seeds are collected during the summer. The lower female section of the cattail pod produces the seeds. Mature seeds can be mashed into a flour that is rich in protein. If the silky part of the seed mass is bothersome, it can be ignited and carefully burned off while helpfully parching the seeds. The tiny seeds may be softened by boiling.

Outdoor Tip: The down, or fluff, from mature cattail seed heads makes good fire tinder. It can be used for stuffing a pillow or as an insulator for warmth when layered between two pieces of material or stuffed into clothing or socks. The fluff was actually used in the Mae West life jackets for flotation! Even to provide light, cattail seed heads can be dipped into fat and ignited as a torch.

POLLEN:

Cattail pollen is produced during early spring. The seed head is divided into two parts with the male portion located above the female. To process, rub, strip, or shake the yellow male pollen into a bag. This substance is sweet and makes a good protein booster but it has a musty flavor and renders pastry products yellow. Sift the fine powder and use with wheat flour in breads and pancakes or alone as a thickener in soups. The pollen can be eaten raw or boiled in water for one-half hour until it thickens into a hot cereal. To store, keep the pollen in the freezer for several months, or it can be dried for future use.

Survival Tip: Protein-rich cattail pollen can be eaten raw–this is an important advantage if there are no cooking means available.

Chickweed

Stellaria spp.

Chickweed

IDENTIFICATION:

We will discuss both common chickweed and mouse-ear chickweed.

Common Chickweed (*Stellaria media*)–The leaves are oval, 1/4" to 1/2" in size, with slightly pointed tips. From each node arise two opposite leaves. Chickweed leaves are green to pale green depending on growing conditions. The plant is low growing and reaches only several inches high; it is frail looking and has a spreading nature.

Multiple, branched stems grow to twelve inches in length. The white flowers are 1/8" to 1/4" in diameter and consist of five petals deeply notched, making them appear to be ten petals. Chickweed is one of the few greens that thrive in mild winters. You will find the best chickweed greens from late fall through early spring. In warm climates the summer heat kills the plant unless it is in a protected area. In northern areas, it can grow all summer. Foragers must remind themselves not to collect the often plentiful chickweed from lawns that have been sprayed with weed killer.

Mouse-ear Chickweed *(Cerastium semidecandrum)* – The leaves are similar to those of common chickweed, but they are less pointed, thicker, and downy or finely hairy.

Caution: Pimpernel (*Anagallis arvensis)* is poisonous. This small annual plant is similar to chickweed when not blooming. It is only a few inches high with opposite leaves. Its flowers are five-parted and may be crimson or blue in color. It is native to Europe but has been introduced into the United States.

In most areas I have traveled, common chickweed is more prevalent than mouse-ear chickweed. It was mouse-ear chickweed that I found first. I was traveling near Charleston, South Carolina, when I decided to take a break from driving. The field was covered with wild mustard that I had identified just the week before. As I walked along the field, I noticed a low, crawling plant that looked to me like the common chickweed identified in the one field guide I had; but this plant was thicker and more hairy. Months later, after truly identifying common chickweed, I realized that I had found mouse-ear chickweed in Charleston. After tasting mouse-ear chickweed, I understood why there is not much written about it: the hairy leaves were not my choice for an appetizing meal! K.L.

RANGE AND ENVIRONMENT:

Chickweed is found widely in disturbed soils, on lawns, in fields, and in shady or damp areas.

IDENTIFICATION BRIEFS:

Species:	*Stellaria spp.* and *Cerastium spp.*
Other Names:	Starwort and Stickwort
Plant Size:	Trailing, 1-4 inches tall
Flower Color:	White

Blooms:	Fall to spring
Sun Required:	Semi-shade
Propagation:	Seed
Plant Type:	Annual

LEAVES:

Gather the leaves late fall to spring and use as a good, bland green. Mix chickweed with other stronger-flavored edibles such as bittercress or wild mustard. Any part of the whole plant can be used raw or cooked. As the

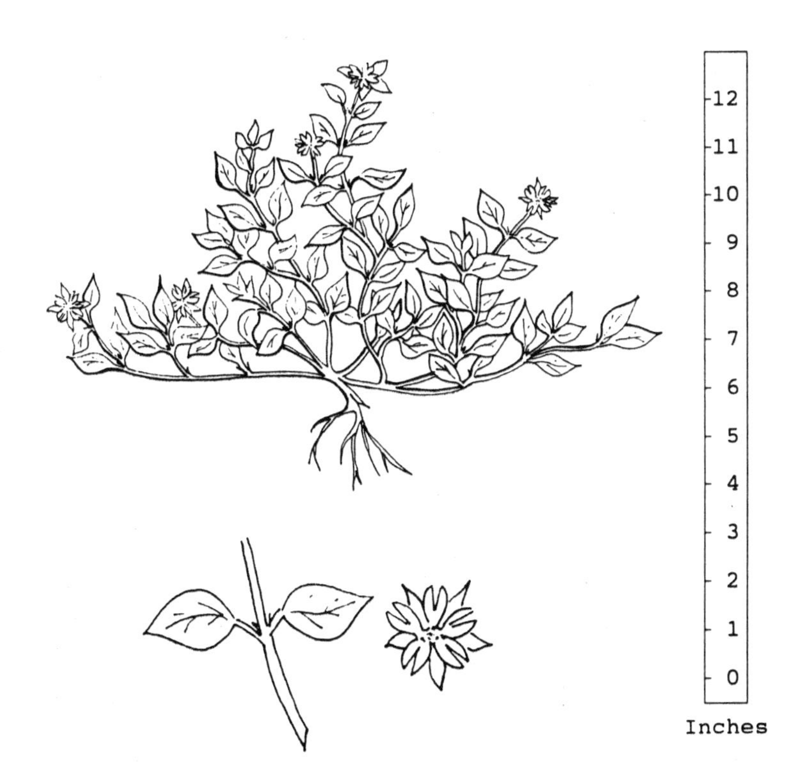

Common Chickweed

Chicory

Chichorium intybus

plant ages, one wants to use only the stem tips and young leaves to avoid stringiness. Chickweed requires only a few minutes to cook, so when combining it with other greens, add it during the last few minutes. When cooked in stews, use a blender to chop.

To freeze, blanch for two minutes, drain, pat dry, and place in plastic bags.

Survival Tip: Chickweed is found during most of the year and has a high vitamin C content. In the past it has been valuable for preventing scurvy in troubled times.

TUBERS:

The roots creep and grow small tubers along the stem. The tubers can be found most of the year and are good raw or cooked.

Chicory

IDENTIFICATION:

The chicory leaf is jagged, deeply scalloped, pointed, and stemless. It is similar to a narrow dandelion leaf, but is coarser and rougher in texture. As the leaves reach the top of the stiff stalk, they turn more lance-shaped and almost toothless. The stalk zigzags upward 1-3 feet with sparse leaves. Flowers are 1 to 1 1/2 inches in diameter and are blue/white in color. Chicory is one of the first greens to appear in the spring and is a good source of vitamin A and C, calcium, phosphorous, iron, and potassium plus trace minerals.

Chicory is unique in having petals that normally close in the noonday sun and are square-tipped while the ends are fringed. Most of the flowers have short stems, but some near the end of the stalk will be longer. The tap root is white and fleshy.

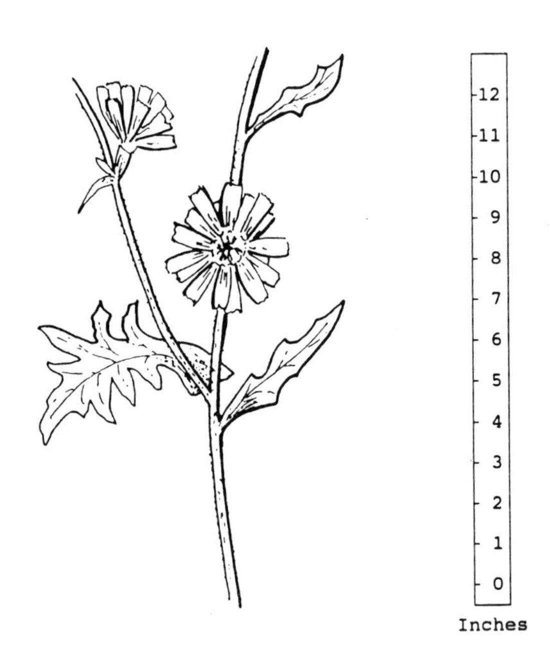

Chicory

I first noticed chicory in Indiana where I was helping victims after a tornado had destroyed a small town. Waiting for a traffic light to change, I noticed a patch of blue wildflowers. I pulled over to get a closer look, and as soon as I saw the fringed square flower petals, I knew it had to be chicory. I had always heard about chicory coffee, and now it was exciting to actually find the plant itself. K.L.

RANGE AND ENVIRONMENT:

Chicory is found widely along roadsides and in other disturbed soils, but it does not like acid soil. Since I live in Atlanta with all its pine trees and acid soil, I see chicory only when I travel north.

IDENTIFICATION BRIEFS:

Species:	*Chichorium intybus*
Other Names:	Succory, Blue Sailors, and Wild Endive
Plant Size:	1-3 feet tall
Flower Color:	Blue/white
Blooms:	Summer through fall
Sun Required:	Full sun
Propagation:	Seed, division
Plant Type:	Perennial

FLOWERS:

The flowers can be sprinkled on salads or added to soups.

LEAVES:

The leaves are collected during early spring. The newly sprouted leaves are young, tender, and not bitter. The white underground part of newly sprouted plants is also good. Use as cooked greens or as a salad. As the plant gets older, you may have to drain off the first parboiling water and then cook again until tender. The leaves can be dried and ground into a nutritional powder that may be used to flavor favorite recipes.

Chicory can be frozen or canned. To freeze, parboil for two minutes, drain, pat dry, and place in plastic bags.

ROOTS:

Roots can be dug all year, but are best during fall, winter, and early spring. Since chicory favors sandy soil, pulling up the roots is an easy task. Use the roots as a coffee substitute or as a coffee extender. To prepare, clean the roots (I use an old toothbrush) and cut into small chunks. Next, slowly roast at 300 degrees in a partly open oven to allow steam to escape. Roast until the white of the roots' centers turn brown and brittle but not charred. Finally, grind up the roots and store in a closed container as you would coffee. Chicory has a somewhat bitter taste when used alone. Use about the same measure as you would for regular coffee or mix 50/50 with regular coffee. The older the roots, the stronger the coffee. Store the powder in an airtight container.

Cook early spring roots and eat as a vegetable or cook in a soup or stew. When gathered before the stems shoot up, roots can be dried, ground into a flour, and used for bread. Chicory roots can be sliced thin and dried to make "chicory chews". Store in an airtight container.

Survival Tip: The coffee made from chicory root improves morale.

Clover

IDENTIFICATION:

Clover usually has three leaflets with or without a pale chevron shape on the leaf. Leaves are round to oval in shape, and they range from 1/2 inch to 1 1/2 inches long. The flowers have round heads 1/2 inch in diameter. They can be white, yellow, or red. Red clover is especially high in vitamin A, protein, and mineral content. It also contains the antioxidant tocopherol.

RANGE AND ENVIRONMENT:

Clover is found widely in fields and lawns.

IDENTIFICATION BRIEFS

Species:	*Trifolium spp.*
Plant Size:	4-8 inches tall
Flower Color:	White or red
Blooms:	Spring to summer
Sun Required:	Full sun or semi-shade
Propagation:	Seed
Plant Type:	Annual

Clover

Trifolium spp.

LEAVES:

The leaves are gathered spring to summer. Before the plant flowers, the young leaves can be eaten raw or cooked briefly. Some like to dip the leaves in salt water before eating raw. I do not enjoy eating clover raw. When cooked, clover is rather bland and can be used to tone down other greens. For older leaves, parboiling and discarding the water may be necessary. Dried leaves can be stored and used later for soups and stews or ground into flour. Clover leaves eaten in quantity may cause bloating and even act as a mild emetic.

FLOWERS:

Gather flowers spring to summer. The young flowers, just before they open, can be eaten raw, perhaps being first dipped in salt water. Some varieties are not very sweet, and parboiling and discarding the water may be necessary. Dried flowers can be used to make a tea or crushed and saved in air-tight jars. For tea, use about one teaspoon per cup of boiling water. Mature flowers and seeds can be ground together into a nutritious flour for bread.

Outdoor Tip: In Ireland, during famines, the dried flowers of clover were mixed with flour as an "extender" in making bread. The same technique could be used when camping.

ROOTS:

The roots are collected fall to winter. They are usually scraped and boiled or can be smoked over a fire.

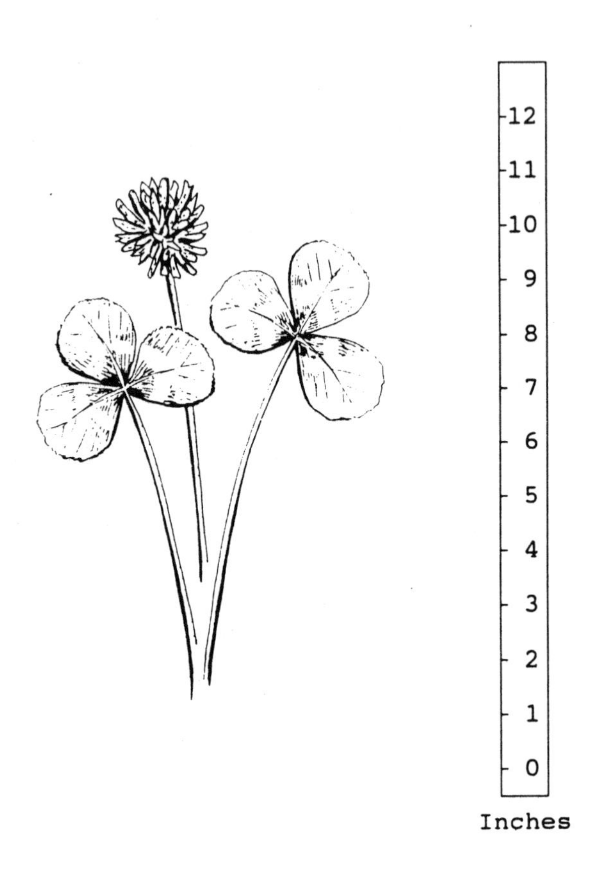

Dandelion

IDENTIFICATION:

Dandelion flowers are yellow/gold. The stems are hollow and the leaves grow in rosettes. The leaves have sharp, irregular, deeply scalloped leaf edges and grow in clusters one foot tall. As the leaves mature, they will exude a white milky sap when torn. This is true for all members of the lettuce family. The taproot is long and large with a yellow skin.

Dandelions outrank most garden vegetables in nutrients with vitamins A and C, iron, copper, potassium, and protein. The deep roots gather many trace elements. Dandelions are still popular in Europe. In the United States, they were used extensively during the Great Depression for food.

In Ohio helping on a flood disaster, I stopped to talk to someone digging dandelions out of his yard. He told me he used dandelion as a spring tonic and that in his ethnic neighborhood, dandelion was in demand. Those he did not want were dug by his neighbors for their use. Unfortunately, I could not afford to fly him home to Atlanta to dig my dandelions. K.L.

RANGE AND ENVIRONMENT:

Dandelions are found throughout in grassy areas, lawns, and waste ground receiving full sunlight.

IDENTIFICATION BRIEFS:

Species:	*Taraxacum officinale*
Plant Size:	Flat up to one foot tall
Flower Color:	Yellow-gold
Blooms:	Spring to summer
Sun Required:	Full sun or semi-shade
Propagation:	Seed
Plant Type:	Perennial

Dandelion

Taraxacum officinale

LEAVES:

The leaves are used late winter to early spring. They are best when young and will become bitter when older, especially after flower bloom. Many like the young leaves fresh in salads, while others prefer to steam or boil for a few minutes and serve like greens with butter and lemon. Try washing the leaves and roots together to make washing them less tedious. To flavor dandelions, add grated garlic, lemon peel, or basil. Or, you can cook dandelion greens, then chill, chop, and serve as a salad with dressing. Also, young leaves mixed with lemon juice, olive oil, and Parmesan cheese makes a tasty salad or chopped and added to pancakes for a unique flavor. When using older greens, parboil, drain, and add second water. The bitterness can be reduced by soaking the leaves in a baking soda solution (one teaspoon of soda for a gallon of water) for one hour. In addition, the first few frosts in the fall will reduce some bitterness in older leaves. Mix dandelions with bland greens like dock, plantain, or chickweed to "pep" them up. Make a tea from 2-3 tablespoons of dried leaves added to one cup of water.

You can avoid the bitterness of summer by covering the plant with an inverted pot for a few days to blanch the leaves.

Although dandelion is more commonly known, some find that sow thistle is a more agreeable food, being less bitter and more tender.

Contact dermatitis may be caused in certain individuals from the milky sap from torn leaves.

The leaves can be frozen: blanch for two minutes, drain, pat dry, and place in plastic bags. Dried leaves can be stored for later use as a good addition to soups, stews or flour products. Cultivated dandelions grow larger and have a milder flavor than the lawn or wild variety.

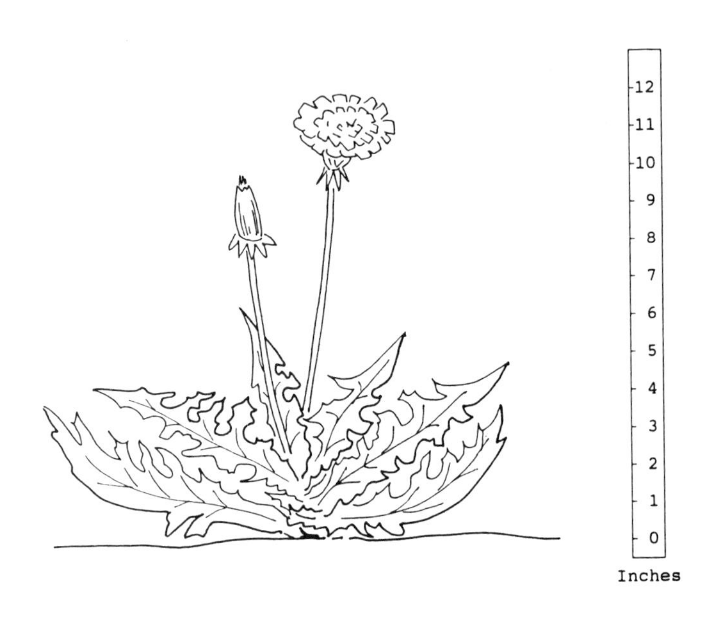

Dandelion

Outdoor Tip: Use the milk from the leaves or stem as an emergency glue.

One year I decided to transplant a particularly stubborn dandelion from my yard to my garden. After treatment like a "real" vegetable, it grew into a monster dandelion that provided many good meals! K.L.

ROOTS:

The roots are collected fall to early spring. They normally are scraped or peeled, sliced, then cooked in boiling salted water and served like carrots. I brush the sliced roots with olive oil, cover with aluminum foil and roast in the oven. During the last minutes, remove the foil for a crisper texture.

For a non-caffeine coffee substitute, medium-size roots are dried very slowly in an oven until dark brown and brittle, but not burned. Medium roots seem to be the best; smaller roots are not as tasty and larger roots tend to be bitter. When the roots cool, crush or grind and store in sealed containers. Use like coffee grounds in a drip coffee maker. The flavor is strong and satisfying. It will not play havoc with your liver (because of the oils) or with your nerves (due to the caffeine.) Use a blend of chicory and dandelion coffee to add variety. Use one teaspoon of ground roots for one cup of coffee. In addition, make a tea by boiling fresh roots in water for 10-15 minutes.

CROWNS:

Dandelion crowns are used during early spring. These whitish crowns are found at and just below ground level and are the tastiest part of the plant. It is the tastiest part of the dandelion. Very mild, they may be eaten steamed, fresh in salads, or coated with a batter and deep-fried. Or the young crown and leaves are excellent wilted in a warm vinaigrette.

FLOWERS & BUDS:

The flowers and buds are gathered during early spring through summer. Dandelion flowers are good and pretty shredded into salads. Both the flowers and buds can be dried to make a tea. To reduce bitterness, be sure not to use any of the bitter stem nor the green calyx at the base of the flower. Typically, dandelion buds are steamed or boiled then covered with butter or cheese. Experiment with cooking the buds in salt water and serve as a vegetable, or pickle it for a special delight. Also the blossoms can be parboiled, drained, and boiled again. Then thicken like peas and serve. They are delicious.

Dock, Curly

IDENTIFICATION:

Curly dock has broad, coarse, lance-shaped leaves with loosely curled edges. Curly dock is a common weed which prefers acid soil. It has many branches with foot-long leaves that are easy to spot. Green seed spikes at the top of the stems form slender clusters which reach 3-4 feet. Later in the season, the seeds turn dark brown.

It was in Duluth, Georgia, that I saw my first curly dock. A corner lot was full of a large, long-leafed plant. Every week I would drive by the lot to casually check out the conspicuous plant's growth. Finally in middle summer, the green seed stalk formed and I knew with no doubt that the plant was curly dock. Now, I can recognize the plant from a distance during any season. K.L.

Curly Dock

Rumex crispus

RANGE AND ENVIRONMENT:

Curly dock is found widely in open fields and pastures and on the roadsides.

IDENTIFICATION BRIEFS:

Species:	*Rumex spp.*
Other Names:	Common Dock, Sour Dock, and Yellow Dock
Plant Size:	3-4 feet tall
Flower Color:	Green, insignificant
Blooms:	Spring to summer
Sun Required:	Full sun
Propagation:	Seed
Plant Type:	Perennial

LEAVES:

The leaves are used in early spring. Tender young leaves are preferred in raw salads because older leaves become tough and bitter. Curly dock is a "user-friendly" wild edible because it produces continuous new growth during the season. In warmer climates, it will have new growth during winter warm spells.

Cook curly dock like other greens, but parboil and discard the water if it is bitter. Minimum water for the second boiling will avoid sogginess. It does not lose its bulk when cooked and is probably best used mixed with wild mustard greens or bittercress and seasoned with vinegar or lemon juice. Although I have never yet tried one, I read that curly dock can be used to make a pie similar to its cultivated cousin, rhubarb.

SEEDS:

Young green seeds may be stripped off the stalk, boiled, drained, then added to soups for a nutrition booster and thickener. The mature seeds are collected during middle summer to autumn. The seeds are edible, but they are small, difficult, and time consuming to separate from their tough, coarse covering though they are plentiful. About the size of sesame seeds, they can be ground into flour then boiled as a cereal or used in bread for extra protein. If the curly dock seeds are used when they turn brown, they become bitter and give bad results. Wait until they turn black and be sure to winnow the seeds from the chaff. Your bread will be black with a

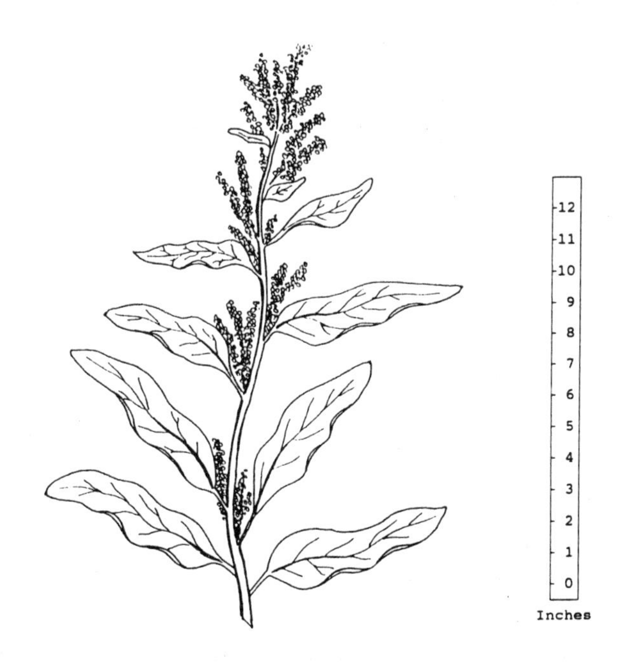

Curly Dock

ROOTS:

The roots may be gathered year long. They are dried, then ground and mixed with other flour.

Evening Primrose

IDENTIFICATION:

Evening primrose is a biennial that has a low round rosette of leaves the first year. The usually singular leaf stalk grows to four feet the second year. Typically, the stalk is reddish and round with some hairs. The leaf is 2-3 inches long, has a stem, and is lance-shaped with a pointed tip. Primrose leaf edges are wavy and toothless and appear in an alternate pattern.

To find the new first-year leaf rosettes, look among last year's dead stalks. The flower blossoms are fragrant, broad petaled, and usually yellow though they are sometimes pink or red in some species. Flowers bloom in the summer and normally open after dark and remain open on cloudy days. A cross-shaped stigma in the center of the flower is a key to identification. The tubular flower buds are left on the stalk after the flower dies.

When I first began searching for evening primrose, I knew it looked similar to goldenrod in growth habit. Then I finally discovered it during the flowering stage when the resulting tubular seed case develops. There was no question it was evening primrose. K.L.

Evening Primrose

Oenothera biennis

RANGE AND ENVIRONMENT:

Evening primrose is found in sparse concentrations throughout the Eastern United States and throughout Britain and Northern Europe, normally along roadsides in wasteground and in sandy, gravely, and sometimes clay areas.

IDENTIFICATION BRIEFS:

Species:	*Oenothera biennis*
Other Names:	Wild Beet
Plant Size:	3-4 feet tall
Flower Color:	Yellow/pinkish red
Blooms:	Summer
Sun Required:	Full sun
Propagation:	Seed, division
Plant Type:	Biennial

LEAVES:

The leaves are gathered during the spring. Young, new rosette leaves of a first year's growth may be added directly to salads for a peppery pick-up. They may be boiled in several changes of water (to reduce the bitter peppery flavor) and served with butter. Some people like to peel back the outer shell of the leaves with a knife, then use the leaves as a spicy addition to salads and other foods. Older leaves have a fuzzy or hairy texture and leave a residue when boiled.

Outdoor Tip: Good cordage can be made by braiding together several strands of crushed evening primrose stalks.

ROOTS:

The roots are collected during the spring. First-year tap roots should be gathered from under rosettes before the plant blooms. (Locate first-year rosettes by looking near old flower stalks.) Scrape or peel the outer layer of roots; slice and boil until tender. Use several changes of water to remove toughness and bitter, peppery taste. Eat as a vegetable with butter or in stews. I like to brush the sliced roots with olive oil, cover with aluminum foil and roast in the oven. Remove the foil during the last few minutes for a crisper texture.

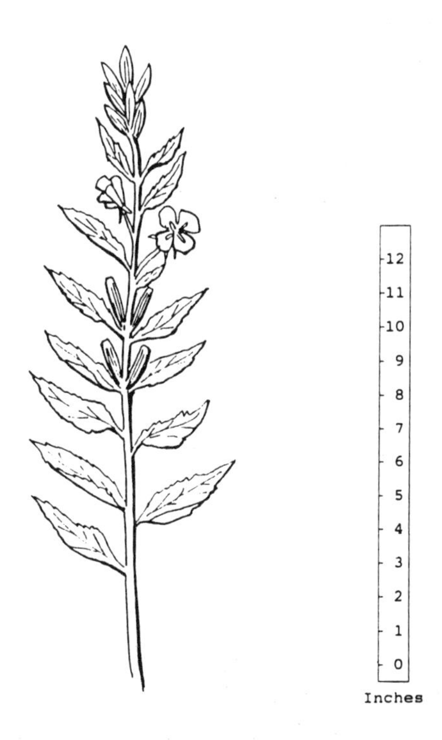

Evening Primrose

SEEDS:

The seeds are found fall through winter. They can be gathered from the tubular seed pods and be ground into flour or meal or boiled into gruel. The seeds are very small when mature but are easy to collect by simply pouring them out of their shell.

When I show others the seed pods, they are always fascinated when the little seeds pour out into their hands, ready to use! K.L.

Goldenrod

IDENTIFICATION:

The sweet goldenrod, *S. odora* species, is normally chosen due to its special, anise-like odor, although all goldenrods are edible. Sweet goldenrod is a frail, slender species growing only two or three feet tall and sometimes reclining, whereas other varieties may grow to six feet tall with rigid, erect stalks. Sweet goldenrod's leaves are smooth on the edges, without the usual toothy jags or ripples of other goldenrods.

All goldenrod leaves are lance-shaped and are generally 3-4 inches long but grow smaller nearer the top of the stem. The stemless leaves are sometimes blotched all over with black, glandular spots. Flowers of sweet goldenrod, the first to bloom in midsummer, are tiny, yellow, and bushy; but they are not as dense as on other species. The seeds are downy.

Goldenrod

Solidago odora

Outdoor Tip: Goldenrod stems can harbor a worm used as fish bait.

RANGE AND ENVIRONMENT:

Goldenrods are found throughout except in Western United States. They prefer dry soil and full sun areas.

IDENTIFICATION BRIEFS:

Species:	*Solidago odora*
Plant Size:	2-3 feet tall
Flower Color:	Yellow
Blooms:	Midsummer
Sun Required:	Full sun
Propagation:	Seed or division
Plant Type:	Perennial

LEAVES:

The leaves are used during early summer through fall. Sweet goldenrod leaves can be used, fresh or dried, to make a pleasant anise-flavored tea. Also, goldenrod leaves may be boiled and used with other greens as a vegetable.

FLOWERS:

The flowers are collected during midsummer and are used to add a peppery flavor to soups.

SEEDS:

The small seeds are harvested during late summer. They can be crushed and added to stews for thickening.

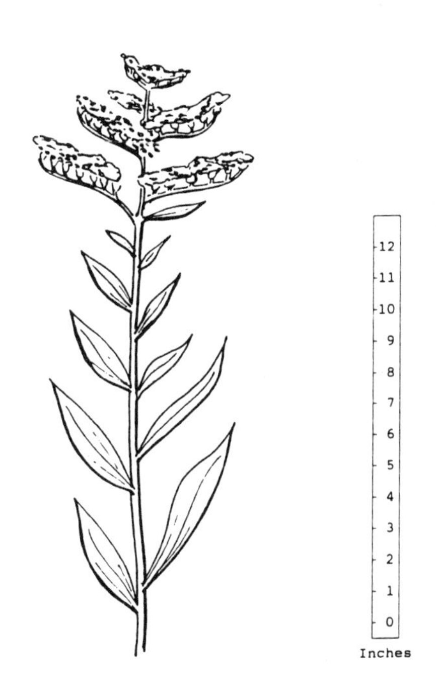

Goldenrod

Grapes

IDENTIFICATION:

Grapes have broad heart-shaped, velvety, simple leaves with saw-toothed edges. The leaves are often lobed (having extensions), the branches have tendrils and can be high-climbing or trailing. The stems have a scaling nature. The small green flowers produce a cluster of round grapes which at maturity range from 1/4 inch to 3/4 inch in diameter and have 1-4 round, pear-shaped seeds.

Warning: Grape roots are poisonous as are the berries of Canada moonseed with which grapes may be confused.

Canada moonseed has no tendrils, and there is only one crescent-shaped seed in its bitter, black berry. The leaf is not attached directly to the leaf stem; instead, the stem intersects under the leaf. The leaf is smooth edged.

RANGE AND ENVIRONMENT:

Grapes are found throughout the United States in moist, fertile soil along the edge of woods, where they can climb.

IDENTIFICATION BRIEFS:

Species:	*Vitis spp.*
Other Names:	Summer Grapes, Fox Grape, Muscadine, and Scuppernong

Plant Size:	Trailing or high climbing
Flower Color:	Green
Blooms:	Early summer
Sun Required:	Full sun or semi-shade
Propagation:	Seed
Plant Type:	Perennial

LEAVES:

The leaves may be used early spring through summer. Young, tender, full-sized leaves can be cooked as a vegetable green, and the very young leaves can be eaten raw. The leaves–boiled for a short time to soften them–can be used to wrap meat or rice for baking. They give a subtle acidic or lemony flavor to the encased foods.

Survival Tip: Military survival manuals say that a drinkable liquid can be obtained from grape vines during the summer. If the vine is cut off near the ground then higher up the vine, sap from the severed vine will drain. I have found that the vine must be in very moist ground or near water to produce a liquid. Use only vines that produce a clear liquid.

TENDRILS:

The tendrils may be gathered during early summer and can be nibbled raw or cooked as a vegetable.

SHOOTS:

The young shoots may be used spring through summer. They can be cooked like asparagus or with other foods to add texture.

Wild Grapes

Vitis spp.

FRUITS:

The fruit is gathered late summer through fall. It is used to make juice or jelly or for eating raw. The rind of muscadines and scuppernongs can be used to make pies. Normally wild grapes are not as sweet as domestic grapes. The young, unripe fruits can be used as a source of pectin.

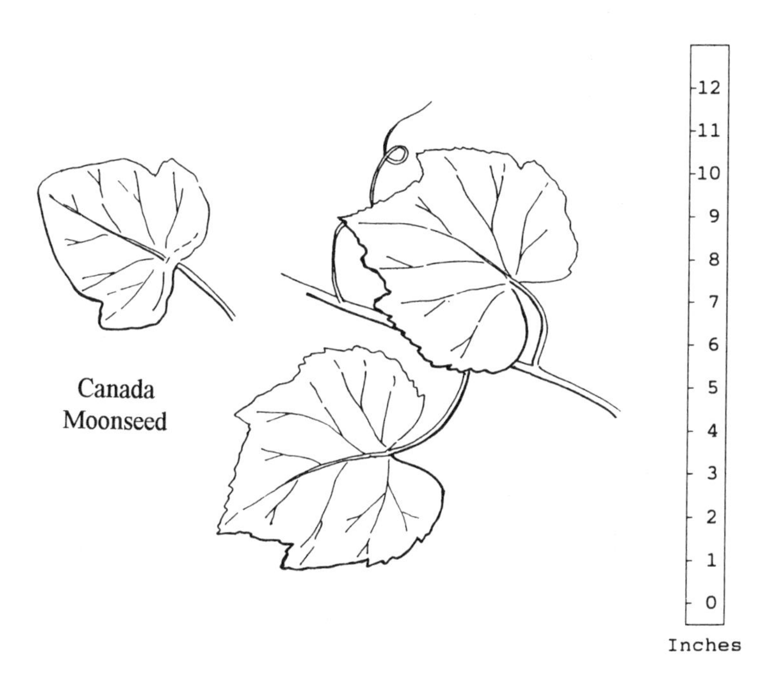

Grapes

Grasses, Bladed

IDENTIFICATION:

Bladed grasses such as rye, wheat, and common lawn varieties are edible as are their seeds. In many areas of the world, grasses make up an important portion of the diet.

RANGE AND ENVIRONMENT:

Grasses are found everywhere in lawns and fields, and along roadsides.

IDENTIFICATION BRIEFS:

Species:	*Graminiae spp.*
Plant Size:	2-3 feet
Flower Color:	Not applicable
Blooms:	Not applicable
Sun Required:	Full sun
Propagation:	Seed
Plant Type:	Annual

LEAVES:

The leaves are gathered all year. Eating grass leaves is not new for powdered grass leaves are sold as a vitamin supplement in health food stores. During the growing season, the leaves plus the white portions of stems, the roots, and tender shoots can be eaten raw or cooked as greens. It is believed that young grass before it joints has the best quality nutrition. Grasses have protein,

Grasses

Graminiae spp.

vitamins, minerals, chlorophyll, and carotene. USDA says Orchard Grass has approximately two times more protein than Timothy. Kentucky Bluegrass was high on nutrients whereas Dallisgrass, Orchard Grass, and Red Top were in the intermediate group. Timothy ranked low class.

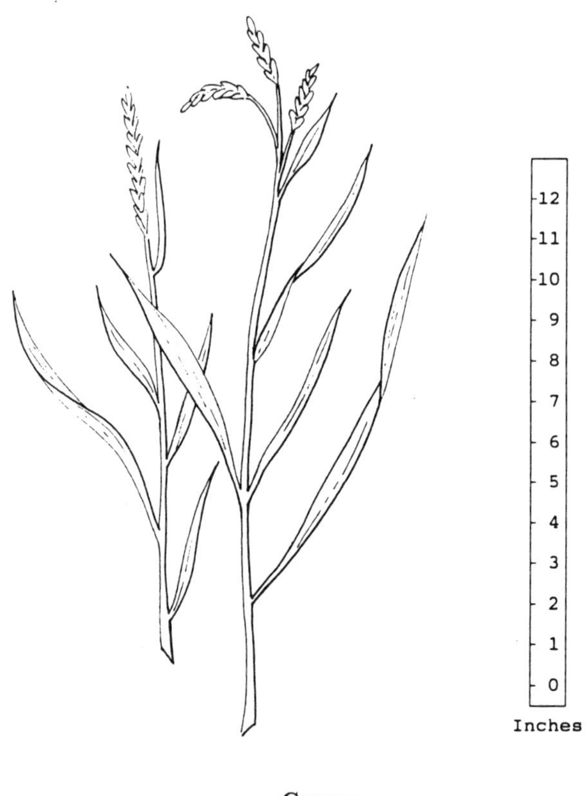

Grasses

One should collect several varieties of grass and taste each group, then collect the most pleasing ones (Timothy and fescue can be bitter). Grasses may be chopped for use in a cream soup. Mature, fibrous leaves may be cooked in liquid to make a nutritious beverage after the leaves are discarded. Dried leaves can be added to milk, soups or ground up and mixed with other flour as a protein, vitamin, and mineral supplement. One researcher indicated that mixing dried grass powder to the daily diet could cut the food bill 25%.

For those who like bean sprouts, I have an interesting substitute. During the winter when greens are scarce, I plant rye or wheat grass seeds in a planter next to a window. When the new growth reaches several inches in length, I clip the tender grass with scissors and use the clippings in a salad. In several more days, I have another fresh crop. K.L.

Survival Tip: Grass leaves contain many vitamins. Tender young grass has been a primary food during many famines. When older, tougher leaves are eaten raw, they should be chewed to extract the juice and the pulp spit out.

SEEDS:

The seeds are gathered summer to fall. Raw seeds are normally safe, but some wild grass seed varieties can have a toxic fungus. It is, therefore, wise to boil or roast all wild varieties. No seeds should be used that have a blackish tint which may represent a fungus. Roasted grass seeds may be ground into flour or cooked whole in water to make a mush. Seeds can be added to soup or bread.

Different seeds have different flavors. When selecting in the field, one should look for larger seeds. Domesticated grains are much easier to work with than wild seeds. Seeds can be sprouted as mentioned before but some wild seeds can take up to a month to sprout.

Outdoor Tip: Grass seeds are rich in protein. If you find a concentration of weed seeds, use them. In urban areas, look for vacant ground that has been seeded with rye or wheat which goes to seed when not cut. They are easy to harvest and process.

Greenbrier

IDENTIFICATION:

Greenbrier has thorny semi-woody, climbing vines with tendrils. The variformed leaves tend to be round or oval in shape. They are leathery, with the veins running parallel. In temperate areas, one will find the plant still green in the winter. Its fruit is small, black with a bluish tint, and grouped in little clusters. Two of the most well-known varieties are common greenbrier and bullbrier greenbrier. Common greenbrier has broad heart-shaped leaves and very thorny stems. Bullbrier greenbrier has triangular heart-shaped leaves, larger shoots and roots, and stems that are less thorny.

I found greenbrier entirely by accident. While at work I was complaining about a plant invading my woods at home and how its thorns would tear skin and clothes. A good friend said I must have a crop of greenbrier. The name had a familiar sound, so that afternoon, I did some research. Sure enough, it was

Greenbrier

Graminiae spp.

greenbrier. I now cherish cooking the tasty young tips every spring. If you have ever had your clothing snagged by the thorns of a green, semi-wood vine while in the woods, you have met greenbrier! K.L.

RANGE AND ENVIRONMENT:

Greenbrier is found throughout the United States in open woods and bottom land.

IDENTIFICATION BRIEFS:

Species:	*Graminiae spp.*
Other Names:	Cat-brier, Bullbrier, Devil-Vine, Saw-brier, Chiney-brier, and Smilax
Plant Size:	High-climbing vines
Flower Color:	Green
Blooms:	Summer
Sun Required:	Semi-shade
Propagation:	Seed or division
Plant Type:	Perennial

Outdoor Tip: The larger bullbrier thorns make effective spur attachments for hooks and fish spears.

LEAVES:

The young leaves can be harvested from spring through summer and eaten raw or cooked as a vegetable.

SHOOTS AND TENDRILS:

The young shoots are collected from spring through summer and are eaten raw or cooked like asparagus, with butter. Greenbrier shoots make a nutritious uncooked trail food. When eating raw, bullbrier shoots are the least bitter.

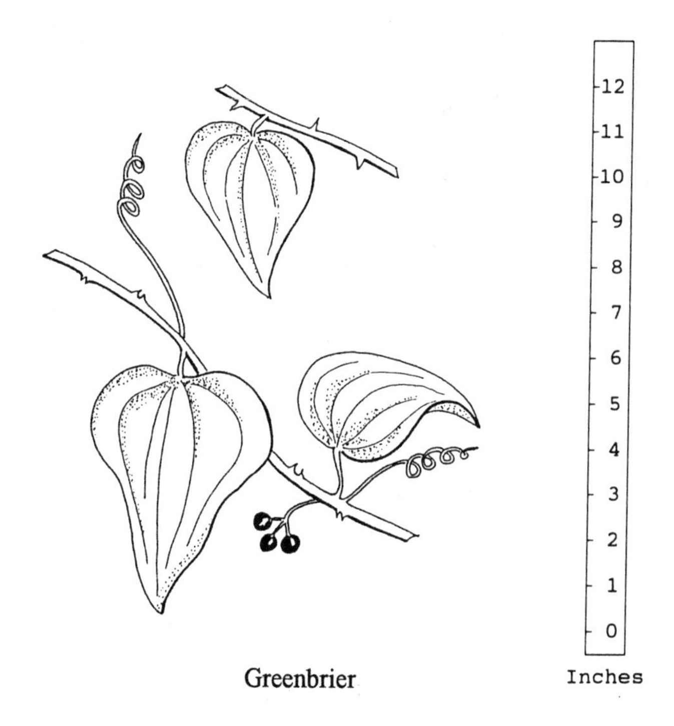

Greenbrier

Outdoor Tip: Good cordage can be made by braiding together several strands from greenbrier stalks, though this cordage becomes brittle when dried. Also, baskets can be woven from the stalks.

ROOTS

The white root can be used all year as a food. The root is crushed and processed as cattail roots. The dried red powder can be used as a thickener, mixed with other flours for natural sweetness, or diluted with water and sugar added for a natural and refreshing drink.

Henbit

IDENTIFICATION:

Henbit is one of the earliest greens to appear in the spring and is easy to recognize because of its unusual appearance. It has a square stem that reaches 6-8 inches in height and ruffled leaves formed into whorls that radiate perpendicularly up the stem at equal intervals. The very delicate flower, purple or pale pink, has a strong lip making it look like a little bugle.

RANGE AND ENVIRONMENT:

Henbit is found throughout in fields, on lawns, and along roadsides.

IDENTIFICATION BRIEFS:

Species:	*Lamium amplexicaule*
Other Names:	Arch Angel, Deadnettle, and Dumb Nettle
Plant Size:	6-8 inches
Flower Color:	Purple to pale pink

Henbit

Lamium amplexicaule

Blooms:	Summer
Sun Required:	Sun
Propagation:	Seed
Plant Type:	Annual

LEAVES:

The leaves may be gathered late winter through spring. They can grow all year but are normally not found in winter nor during hot summers. Henbit can be eaten raw, but it is best chopped in small pieces and cooked in a soup. The young, tender stems, leaves, and flowers are used as a marginal-tasting green. The older plants and flowers become too chewy and seedy for most palates.

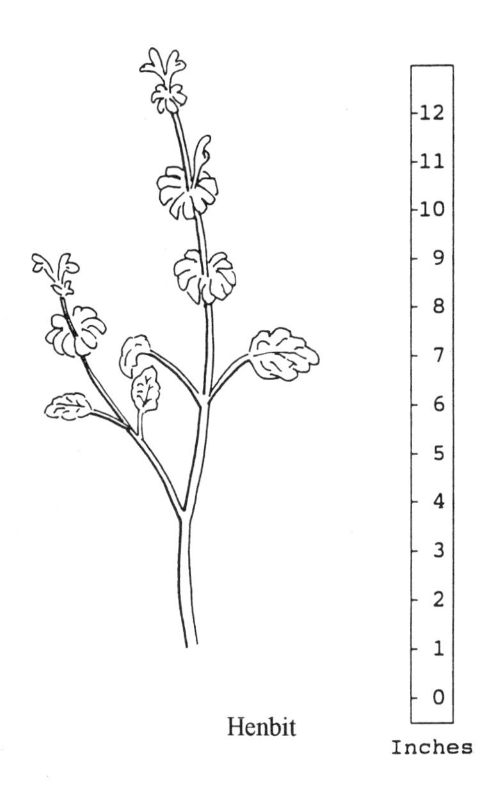

Henbit

Kudzu

Pueraria lobata

Kudzu

IDENTIFICATION:

Kudzu is a long, high-climbing vine with large spade-shaped leaves that will completely cover trees. The stems and bottoms of its leaves are covered with fine hairs. Blossoms have a "pea flower" shape and are purple. Roots range from small to several pounds. Both the vine and leaves are rich in protein. Due to its rapid growth, kudzu was originally used in the South as livestock fodder and erosion control.

RANGE AND ENVIRONMENT:

Kudzu is found in the Southeastern United States in disturbed soils.

IDENTIFICATION BRIEFS:

Species:	*Pueraria lobata*
Plant Size:	High-climbing vine
Flower Color:	Purple
Blooms:	Summer
Sun Required:	Sun or semi-shade
Propagation:	Seed or division
Plant Type:	Perennial

LEAVES:

The leaves are used spring to fall. All but the very tiny new leaves are too tough and fibrous to eat. The young leaves have a hairy texture and should be finely chopped up and cooked like turnip greens. Leaves are marginal in taste and should be used in soups or mixed with other foods.

Kudzu Quiche

4 cups of very small leaves
1/2 cup of milk
2 cups grated cheese
4 eggs
Pie shell

Wash and boil tender leaves in salted water for 15 minutes. Drains and finely chop. Mix with milk and beaten eggs. Place alternate layers of cheese and kudzu mix in pie shell. Bake at 350 degrees for 30-45 minutes until set. Check by sticking a toothpick in the center. If it comes out clean, the quiche is done. Allow to cool for 10 minutes before serving.

Survival Tip: In a survival situation, the older, coarser leaves can be chewed. The juice is swallowed to get the nutrients, and the pulp is spit out.

FLOWERS:

The flowers are collected during the summer and are eaten raw or added to other foods.

KUDZU JELLY

4 cups kudzu flowers
1 teaspoon lemon extract
1 package pectin
5 cups sugar

Chop kudzu flowers and place in a bowl. Pour one quart of boiling water over the flowers and set in refrigerator overnight. The next morning, strain and add one teaspoon of lemon extract and one package of pectin. Bring the mixture to a full boil. Add sugar, then bring to second boil, stirring constantly. Allow to boil for one minute. Skim, then pour into sterilized jars and seal. Same recipe can be used for many types of flowers, including roses, violets, wild carrot, and dandelions.

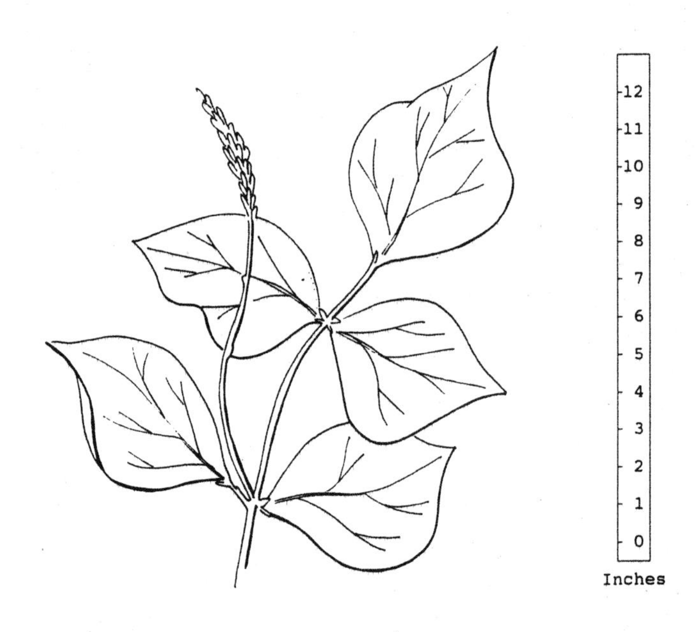

Kudzu

SHOOTS:

Shoots are gathered spring through fall during kudzu's continual state of rapid growth. Only the first four-or-so inches of new shoots (they will break rather than bend) should be used. These sections will be tender but chewy and a little hairy even when cooked. The outer skin can be carefully removed, leaving a tasty, asparagus-like core. Chopped-up shoots, one inch long, can be boiled in salt water with bacon and cooked like string beans. They can also be sautéed, but kudzu is best combined with other foods.

Survival Tip: The part of the shoot that only bends and will not break has a tough outer layer when cooked. The soft core can be sucked out, or you can chew the shoot to obtain nutrients and spit out the pulp. Kudzu vines can be stripped down into usable fibers for cordage.

ROOTS:

The roots are collected all year, but winter kudzu roots offer the highest concentration of starch. In Asia, only the main taproot is eaten, but that can be difficult to find. Most of the other roots are fibrous. After the outer bark is removed, the tap roots can be boiled and eaten.

Oriental and health food stores market a commercial cooking starch called "kuzu" made from kudzu root. The starch is made by washing it out of the root as is done with cattail roots and used as a soup thickener. To be stored, the starch can be dried in the sun or at a low temperature in the oven with the door ajar to allow the water vapor to escape.

Survival Tip: The roots are very fibrous but can be chewed to obtain the carbohydrates and the pulp spit out. If heat is available, cooking the roots makes the carbohydrates more digestible. Process roots as shown in the cattail section.

Lamb's-Quarters

IDENTIFICATION:

Lamb's-quarters, like amaranth, is sometimes called pigweed because it is found in rich soil around pigpens, and pigs like it. It is also called goosefoot, as the shape of its leaves resembles a goose's foot. Lamb's-quarters can grow to six feet tall in good soil. It is an erect plant with many branches. Its stems are straight with a ridged or angular shape.

When mature, the stems can be red or purple-streaked. Upper young leaves are narrow and lance-shaped, with smooth edges. The mature leaves have a triangular shape, and the bottom and top of the leaves have a gray-green-whitish, powdery look. When dipped in water, the leaves will not get wet. Seeds are small and black. The plants like a rich alkaline soil.

Outdoor Tip: Lamb's quarters' roots make a good soap substitute. Just wet your hands and scrub with the crushed root piece.

RANGE AND ENVIRONMENT:

Lamb's-quarters is found throughout in fertile plowed ground, fields, pastures, and in other places of rich soil.

Lamb's-quarters

Chenopodium spp.

IDENTIFICATION BRIEFS:

Species:	*Chenopodium spp.*
Other Names:	Goosefoot, Pigweed, Wild Spinach, Poorman's Spinach, and Mutton-tops
Plant Size:	Up to six feet tall
Flower Color:	Greenish and insignificant
Blooms:	Summer
Sun Required:	Sun
Propagation:	Seed
Plant Type:	Annual

LEAVES:

The leaves may be used late spring to summer and are usually gathered later than most greens. Lamb's-quarters is a favorite wild green because of its mild flavor which, when cooked, is similar to spinach in taste and texture. It does not bolt in heat and if under six inches, the whole plant can be used, or the tender top and side leaves may be used as the plant ages. Lamb's-quarters leaves can be dried, crushed, and used as a thickener. They can be canned, using normal procedures, or excellent frozen. To freeze, blanch for two minutes or steam for one minute, pat dry, and place in plastic bags. Gather a large quantity because lamb's-quarters will lose bulk when cooked. A good source of beta-carotene.

Warning: If the crushed leaves of a plant smell like varnish, do not eat it as you probably have found the wrong plant! Lamb's-quarters resembles the Mexican tea plant, but Mexican tea has leaves with more wavy edges than the more pointed leaves of lamb's-quarters.

When I was first trying to find lamb's-quarters, I was very frustrated. Everything I read referred to it as being readily available and excellent tasting, but I never saw any in the poor soils around home. All the pictures I had seen showed a delicate looking plant, small in size.

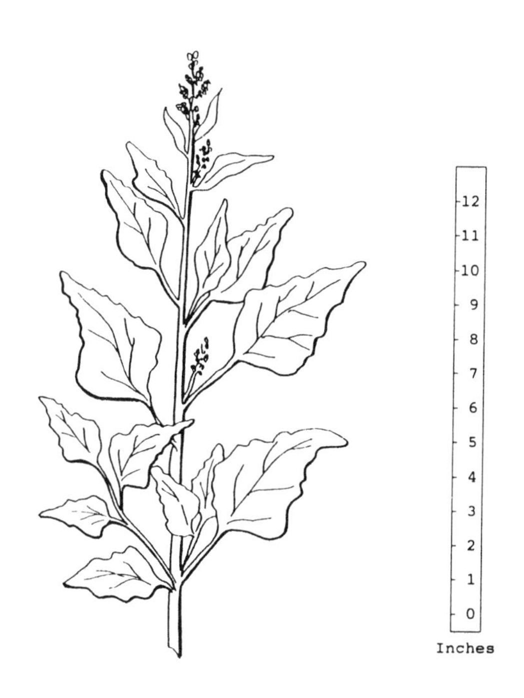

Lamb's-quarters

Then one day at work I parked next to a vacant lot with rich top soil. There was this large seven-feet tall by five-feet wide plant shading my car. I noticed it had a funny looking maple leaf shape with a chalky underside. My mind suddenly recalled the characteristics of lamb's-quarters. To add insult to injury was another discovery the next year after I put several loads of manure on my garden: soon a crop of young lamb's-quarters was growing there and, yes, the young plants looked just like the drawing I had studied. K.L.

SEEDS:

The seeds are collected late summer to early winter–they stay on the plant well into frost. Seed harvest is simple. After the seeds are formed, one may hang the plant upside down over a newspaper or cloth which will catch the seeds as they mature and fall. Or one may dry the plant, place it into a bag, shake the bag, then remove and discard the stalk to leave a package of seeds.

The black seeds can be used to top muffins and other breads, cook as hot cereal, or make into meal or flour. Most varieties I have found have seeds that are not suitable for grinding because they are too small and slippery. They can, however, be boiled for several hours and then crushed into mush which is dried for flour. Some people strip the leaves and seeds off the plant then dry and grind them into flour.

Studies have shown that seeds are a good source of protein and carbohydrates rated higher than corn.

Prickly Lettuce

Lactuca spp.

Lettuce, Wild

IDENTIFICATION:

Two types of wild lettuce are considered here: prickly lettuce and wild lettuce.

Prickly Lettuce–This plant has lettuce-type leaves of two basic types. One is deeply scalloped, and the other is bluntly lance-shaped. Both types show a row of stiff spines on the leaf which grow 6-10 inches in length. It is sometimes called a compass plant because the leaves tend to follow the sun. Older stems and leaves give off a bitter milky sap when cut or torn.

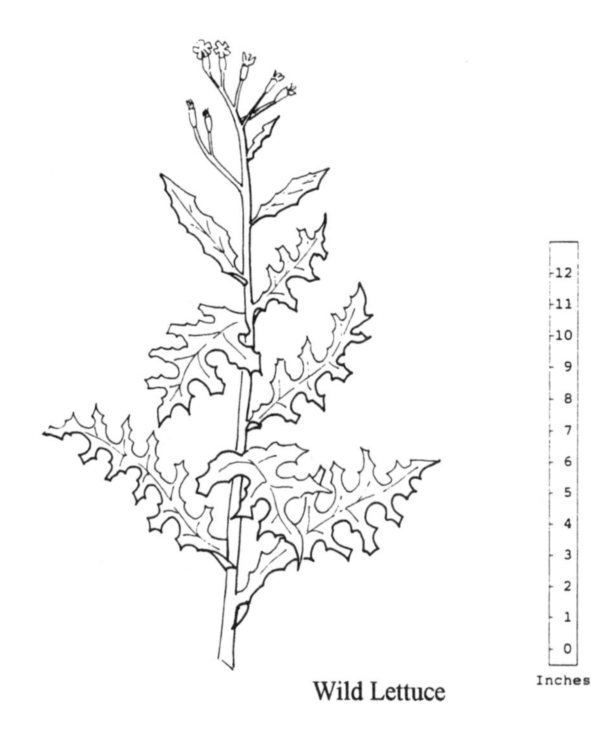

Wild Lettuce

Wild Lettuce–Wild lettuce includes several different species with variable leaf patterns. They range from smooth garden-type leaves to deeply scalloped leaves with pointed tips. They grow to ten inches in length. Stems are smooth. The leaves give off a milky sap when cut or torn..

RANGE AND ENVIRONMENT:

Wild lettuce is found throughout in rich soils of roadsides, clearings, and fields.

IDENTIFICATION BRIEFS:

Species:	*Lactuca spp.*
Other Names:	Compass Plant
Plant Size:	Prickly Lettuce–2-6 feet tall; Wild Lettuce–1-3 feet tall
Flower Color:	Yellow
Blooms:	Summer
Sun Required:	Full sun
Propagation:	Seed
Plant Type:	Annual

LEAVES:

The leaves are gathered early spring to summer. On younger prickly lettuce leaves, the spines are soft and the leaves can be eaten raw. Older leaves lose their prickly nature when boiled.

Wild lettuce has a bitter flavor similar to that of dandelion. It is normally not eaten raw due to the bitterness and the upset stomach it can cause when eaten in large quantities.

One may parboil it a few minutes, drain it, add new water, and cook it until tender (unfortunately, parboiling does increase the loss of vitamins). Wild lettuce should be used while it is under one foot in height, and it does not freeze well.

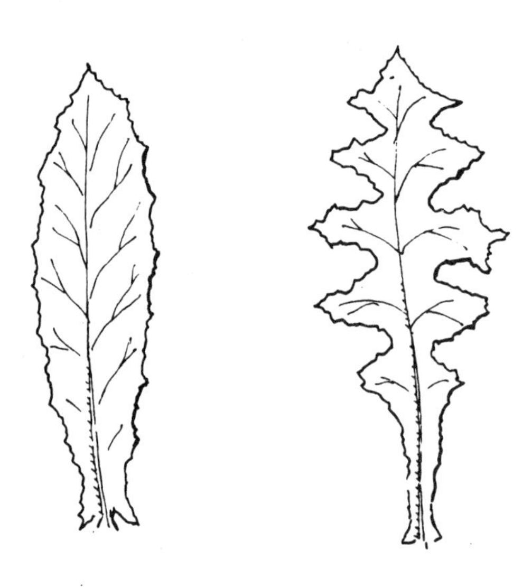

Various shapes of prickly lettuce

FLOWER HEADS:

The flower heads can be eaten before they open and blooming takes place during the summer. The flower

Maple Tree

Acer spp.

heads offer a unique flavor and can be used in soups, baked in casseroles, or fried like fritters.

Maple Tree

IDENTIFICATION:

Maple trees grow to 60 feet in variable dome shapes. The leaves have three or more long pointed lobes that are toothed. The seeds are doubled and winged. There are many varieties with sugar maple being the preferred variety.

RANGE AND ENVIRONMENT:

Maple trees are found anywhere throughout the United States, but they prefer damp woods.

IDENTIFICATION BRIEFS:

Species:	*Acer spp.*
Plant Size:	Up to 60 feet tall
Flower Color:	Not applicable
Blooms:	Not applicable
Sun Required:	Full sun or semi-shade
Propagation:	Seed
Plant Type:	Perennial

SEEDS:

The small green unripe seeds are collected during the fall. They are shelled and roasted or boiled until tender and eaten like peas.

TREE BUDS/LEAVES/SEEDLINGS:

The tree buds or newly opened leaves are an excellent source of trace minerals. They can be nibbled fresh or added to salads. The young seedlings from newly germinated seeds are gathered during early spring. They

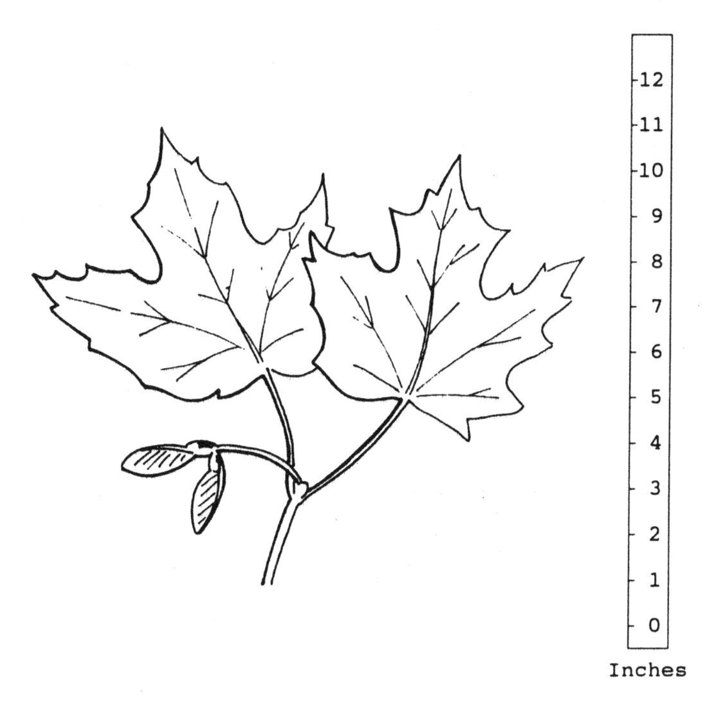

Maple Tree

are added near the end of the cooking process with other vegetables. Young seedlings can also be chopped and eaten raw in salads or dried for future use.

INNER BARK:

The inner bark is used during the spring for eating either raw or cooked.

Outdoor Tip: A low-grade flour can be made from the maple tree inner bark cambium layer. The bark is air dried and ground into flour. Ash cakes can be made from the flour mixed with water and cooked right in the hot ashes of the fire.

SAP:

The sap is tapped from late January until trees begin to bud. The best times for tapping are those when the days are warm and the nights are still cold or even freezing. During the day, the sap begins to move toward the leaves; and during the cooling nights, the sap moves down. The sap travels up to three inches beneath the bark and can be used directly from the tree as a sweetened liquid. It is processed through extensive, careful boiling to become maple syrup. Unfortunately, many gallons of sap are needed to form one pint of syrup.

Freezing is another method that can be used to concentrate the sugar. After the freezing, one should thaw the frozen sap until one third of it is melted, and then pour off the liquid. (The sugar solution melts more rapidly than the water. We see the same concept when a quart of apple juice is frozen and the juice melts first, leaving only the ice core.) The sap can be frozen again to further reduce the water content. The resultant liquid can be used as a sweetener.

Under good conditions, a single tree can produce a gallon of liquid a day. The sugar content of the various species differs greatly.

To tap a tree, one drills a 1/2 inch hole about waist high. Some harvesters tap on the southern side of the tree,

Milkweed

Asclepias syriaca

believing the greater reception of sun causes the sap to move better. The hole is made two inches deep and pointing slightly upward. It is best to tap either above a large root or below a large branch. A small spout, made from a rolled can lid, carved wood, or hollowed-out sumac or elderberry branch is inserted. The liquid drips into a pail hung by the handle on a nail above the spout. Two spouts may be placed in an average tree. Sap ferments rapidly, so the pails should be emptied daily and the sap frozen for future use or for processing into syrup. When sap is no longer desired or the season is over, the hole is plugged with a wooden peg.

Outdoor Tip: Maples are discussed here but all of the birches, shellbark hickory, black walnut, butternut, and sycamore all can be tapped for sweet and edible sap. Birch and maple are especially good, as their sap yields quick sugar energy.

Milkweed

IDENTIFICATION:

Milkweed is a sun-loving, stout, and normally single-stalked perennial. Leaves are large, opposite, 6-8 inches in length, greenish-purple, velvety-hairy on the lower surface, and wavy-edged. Milkweed grows 3-6 feet tall. When its stems or leaves are broken, a milky sap is given off. The pink or red flowers have five reflex petals which later turn into seed pods that are lumpy and large. Milkweed has a distinct taste requiring that the plant be cooked before use.

Caution: The toxic plant, dogbane, also has a milky sap, but it has smaller, smooth leaves, no marginal vein, reddish stems, and is bitter and unpalatable whereas milkweed has a downy or hairy covering and marginal veins connecting the tips of the primary veins. Dogbane's seed pod is long, thin, and pencil like. Milkweed has a seed pod with a fat, swollen base and a tapered tip.

RANGE AND ENVIRONMENT:

Milkweed is found in Eastern and Central United States in sweet, dry disturbed soils and along roadsides. It is sparse in the South.

IDENTIFICATION BRIEFS:

Species:	*Asclepias spp.*
Other Names:	Butterfly-weed and Silk-weed
Plant Size:	3-6 feet tall
Flower Color:	Pink to red
Blooms:	Summer
Sun Required:	Full sun
Propagation:	Seed
Plant Type:	Perennial

LEAVES:

The leaves are gathered during the spring when they are first opened. They should be parboiled three minutes, then the bitter water discarded and replaced with clean boiling water. (Cold water tends to fix bitterness). One should repeat this process three times, then cook the leaves for 15 minutes before seasoning them. A pinch of soda can be added during cooking to break down the fiber and improve flavor.

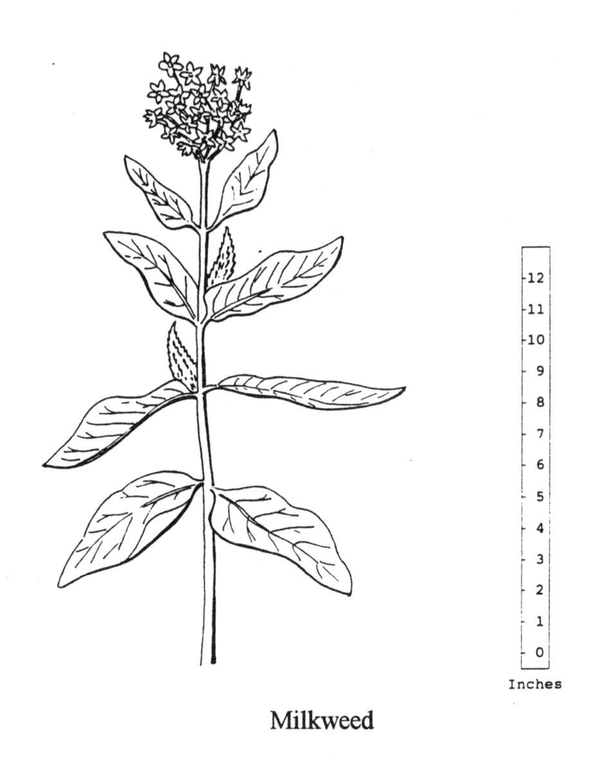

Milkweed

SHOOTS:

The young shoots under six inches long, found during the spring, are used as a vegetable. The fuzz on the shoot is removed by rubbing it off. Preparation is the same as for the leaves.

Survival Tip: The fiber from the stalk can be used to make cordage, coarse weaving material, and fishing lines. The fibers, even when wet, insulate as well as goose down.

Outdoor Tip: Milkweed fiber from pounded dead stalks makes good fire tinder as does the silk from mature pods.

FLOWER BUDS AND FLOWERS:

The flower buds and flowers are collected during the summer. The flower buds are dipped in boiling water for one minute, battered, and fried like fritters. Some substitute the flower buds for broccoli and when cooked, they are mucilaginous–similar to okra. The flower clusters may be battered and fried. Buds and flowers as well as the leaves can be frozen after being cooked.

SEED PODS:

The seed pods, no larger than a walnut, are gathered during the summer and used, like okra, in soups. Also like okra, pods must be gathered at the right time before they become tough and elastic. A bit of baking soda in the water will help break down the tough fibers in the seed pod. Parboiled for several minutes, the young pods may be slit, rolled in a cornmeal/flour mixture, and fried or frozen for future use.

Mustard, Wild

IDENTIFICATION:

Wild mustard is an annual plant. Its lower leaves have stalks that are deeply scalloped and broad, but as the leaves reach the upper ends of the seed stalk, they are small, toothed, and not scalloped. The seed pods point skyward. They are slender and tapered to a point. There are two types of wild mustard:

Field Mustard–This more common wild mustard has a blunt-ended, crinkled leaf.

Black Mustard–Black mustard has a pointed, somewhat bristly leaf.

Flowers of both field and black mustard are made up of four yellow petals. The flower stalk may reach two to three feet in height.

The pods are filled with dark seeds which when dried and ground produce a black pepper equivalent.

RANGE AND ENVIRONMENT:

Wild mustard is found throughout in old fields, vacant lots, or anywhere else there is disturbed ground.

IDENTIFICATION BRIEFS:

Species:	*Brassica spp.*
Other Names:	Field Mustard and Black Mustard

Wild Mustard

Brassica spp.

Plant Size:	Up to three feet tall
Flower Color:	Yellow
Blooms:	Summer to fall
Sun Required:	Full sun
Propagation:	Seed
Plant Type:	Annual

LEAVES:

The leaves are used in early spring before the plant flowers. Young tender leaves can be used like other fresh salad greens, or as most prefer them, cooked the same as domestic mustard greens. Chopped and used to spice up salads, the young leaves add the characteristic pungent mustard smell and taste. They need the routine of

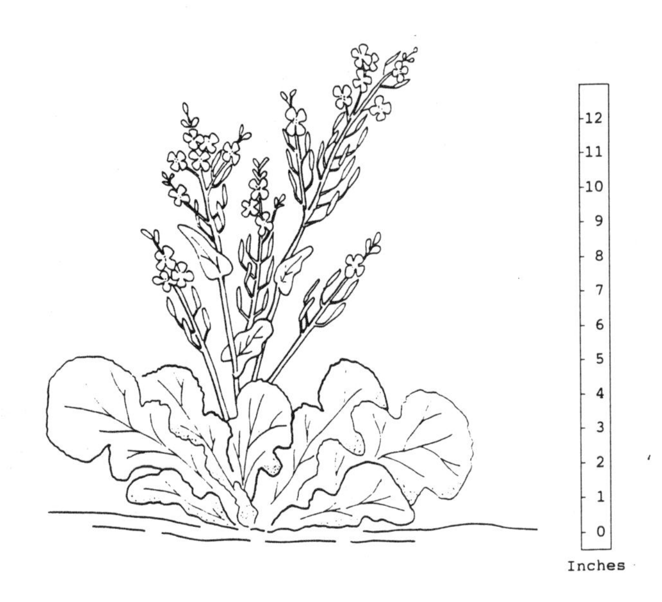

Wild Mustard

Oak Tree

Quercus spp.

parboiling (10 minutes), rinsing, and then cooking until tender. Mustard greens require more cooking than many greens and lose their bulk when cooked.

SEEDS AND SEED PODS:

The seeds and pods are collected summer to winter. They are used to garnish salads or add to soup. Mature seeds can be dried and ground for pepper, but additional drying time should be allowed before storage. To make the condiment mustard, mix the ground seeds with milk or vinegar.

FLOWER BUDS AND FLOWERS:

Flower buds and flowers are found summer to fall. The unopened flower buds and the flowers are cooked for only a few minutes. One should not mix them with the upper stem leaves that are very bitter. The buds are a lot like broccoli when served with butter.

Oak Tree

IDENTIFICATION:

There are many species of oak trees. Normally, they are divided into two major groups:

Red Oak–The red oaks have deeply scalloped leaves with very pointed tips. The acorns from the red oak are very bitter. The acorns require two growing seasons to mature, have a hairy lining on the inside of the shell, and the nutmeats are yellow in color. Red oaks are also members of the black oak family.

White Oak–The white oak also has leaves with deep scallops, but the tips are rounded. The acorns of the white oak are less bitter than those of the red oak, and they require only one growing season. The inner portion of the white oak acorn shell is smooth, and the nutmeat is white in color. The chestnut oak is considered part of the white oak classification.

RANGE AND ENVIRONMENT:

Oak trees are found throughout. They prefer open woods and bottom land.

IDENTIFICATION BRIEFS:

Species:	*Quercus spp.*
Other Names:	Red Oak, Black Oak, White Oak, Chestnut Oak, and others
Plant Size:	Up to 70 feet tall
Fower Color:	Not applicable
Blooms:	Not applicable
Sun Required:	Full sun or semi-shade
Propagation:	Nuts
Plant Type:	Perennial

Outdoor Tip: A tannin solution can be used as soapy water. The bark of oak and most other hardwood trees will provide a solution of tannin by boiling it gently for an hour or two.

NUTS:

The nuts are gathered during the fall from September to October. When processed properly, acorns have a pleasant nutty flavor. Acorns are an excellent source of energy, protein, carbohydrate, and calcium.

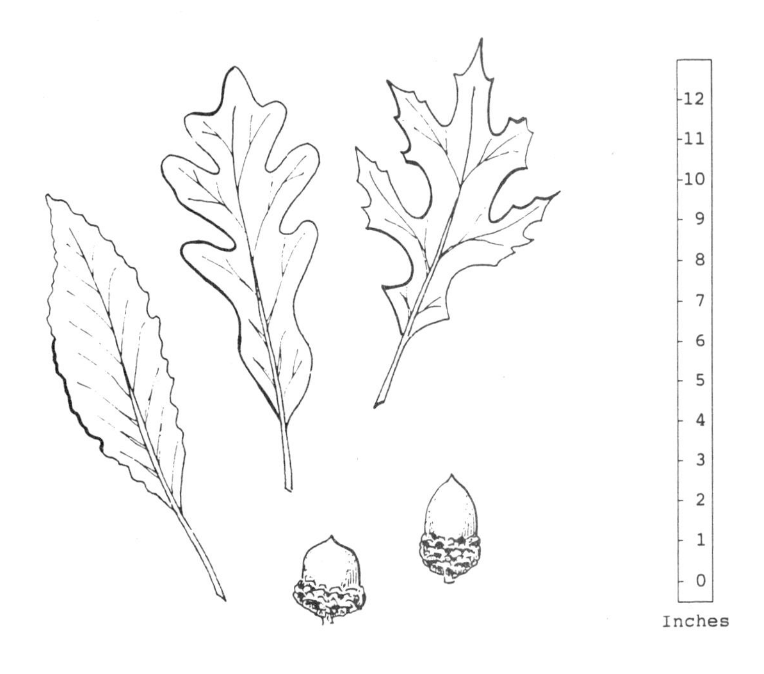

Oak Tree

When collecting acorns, one should not be surprised that many of them must be discarded due to insects or mold, so more should be collected than are needed. If you spread a sheet of plastic under the tree and use only those acorns that fall within a one-day period, this seems to reduce bug infestation, an especially important problem for acorns that are to be stored in their shell. The ripe tan-to-brown acorns, rather than the unripe green ones, should be gathered.

The bitterness in acorns is caused by tannic acid which is water soluble. To remove this unpleasant taste, shell the brown, ripe acorns and remove any corky skin layers, dice the meat; and boil the chunks in water from 15 to 30 minutes until the water turns brown. Then pour off the water and repeat the process until the water clears, indicating that the tannic acid has been removed. Periodically taste a bit of the acorns until you no longer detect any bitterness.

Native Americans would let the crushed acorn meat soak in a fast-moving, clean stream for several weeks to remove the bitterness.

Another method is to boil them in the shells first. This makes the shells soft and easier to remove. The oily coating on the acorns is removed during the leaching process. During the last boiling, salt water can be added; then the acorns can be deep fried or mixed in a soup. Also, finely chopped acorn meats can be added to bread and muffins, or the soft acorn nut can be added as a protein booster to cooked greens. After the leaching process, acorn meat can be frozen. Unfortunately, boiling results in a loss of oils and flavor.

Another leaching method involves bringing a pan full of shell acorns and water almost to a boil. Then turn off the heat and allow to cool for 24 hours. Pour off the water and add new water and repeat the process one or two more times. This type of "cold" leaching results in a more nutritious, flavorful, and softer acorn to grind when dried.

To make flour, the boiled acorn meat can be split in two and dried by slowly baking in a 200 degree oven with the door cracked to allow moisture to escape. Or, they can be dried in the sun. They are then crushed or ground and used as a thickener or as flour. Another method is to roast the fresh acorns to work well in a grinder or blender. After grinding, the course flour is placed into a cloth bag and boiled to leach out the tannic acid.

Acorn flour can be used alone to make an acorn bread, but it is not very pleasing to most tastes. Acorn flour is more palatable when mixed with wheat flour or corn meal–one part acorn meal mixed with four parts corn meal for corn bread, or one to four parts wheat for bread. For a dark acorn bread, mix one cup each of acorn flour and wheat flour, three teaspoons of baking powder, one teaspoon baking soda, one teaspoon of salt, three tablespoons of honey or sugar. Next add one cup of milk, three tablespoons of oil, and a beaten egg into the mix. Bake for 30 minutes at 300 degrees. For pancakes, add more milk or water.

The acorn meal can also be heated in water to make a nutritious mush. Or add enough water to make a thick batter. Add a dash of salt and sweetener to improve the taste. Allow the batter to stand for an hour (or until thick) then pat into pancakes and cook or twist and bake on an open fire.

The leached acorns, after they are roasted until brittle, can be ground and used as a marginal coffee substitute.

Outdoor Tip: Small cakes can be formed from the acorn flour mixed with water. The "ash cakes" are baked in the hot ashes of the fire or on stones.

Survival Tip: A low-grade flour can be made from the oak tree inner bark cambium layer. Strips must be boiled in two changes of water to remove the tannic acid then air dried and ground into flour, not as good as acorn flour, but available all year. In their shell, the dried acorns will store for a time. Some Native Americans stored acorns for several years in bags buried in boggy areas. In survival and famine situations, calories (fats) will be difficult to obtain. Analysis of the acorn meal has shown it to be 65% carbohydrates, 18% fat, and 6 percent protein.

Wild Onion/Garlic

IDENTIFICATION:

Wild onions and garlic have a stronger flavor than domestic varieties, so smaller quantities are used. The onion smell is a cue for use. Wild onion is less common than wild, or field, garlic. Often after the leaves dry up, the flower head will help one find the below-ground bulb's location. Most of the "onions" we see in our yards are really wild garlics.

Wild onion and garlics are an excellent source of vitamin C, contain fair amounts of vitamin A, and are rich in sulfur, iron, and calcium.

Wild Onion–Wild onion reaches about 12 inches in height and is not as tall as the garlics. It has thin, soft, slender, flattened, and linear leaves. It is found in fields, lawns, and rocky open fields that are dry and sunny. The whole plant has a strong onion flavor. It forms star-shaped, white or rose-colored flowers in a large cluster. The flower stem is arching at the top. The wild onion bulb will separate in layers like a domestic onion and it is normally taller than wide.

Wild Garlic–Wild garlic or meadow garlic is very similar to wild onion, and it has an onion, rather than garlic smell. The thin leaves begin at the base of the plant, are narrow, frail, not hollow, somewhat flattened, and concave. It has only a few flowers when blooming. Wild garlic has a seed stalk that grows 12 inches tall when blooming. A 3-part spathe is attached below the 1 inch in diameter group of bulblets. These top bulbs contain a dozen individual bulbs, growing next to each other and each shaped like a tiny flat clove of garlic. The below ground basil bulbs are not divided into cloves like domestic garlic but are solitary with many growing next to each other.

Field Garlic–Field garlic or crow garlic is very similar to wild garlic, but it has a very strong, lingering garlic odor and taste. It grows to two feet tall and is larger than wild onion. The leaves are tough, slender, round, hollow, and attached part way up the main stem. Flowers and bulblets are more numerous than those of

Field Garlic

Allium spp.

wild garlic. A single spathe encloses the bulblets which are divided into cloves.

The photo is of field garlic. Notice the dried single spathe membrane that encloses the bulblets.

RANGE AND ENVIRONMENT:

Wild onion and garlic are found widely and prefer meadows, disturbed soils, and yards.

IDENTIFICATION BRIEFS:

Species:	*Allium spp.*
Plant Size:	12-24 inches tall
Flower Color:	Pink to white
Blooms:	Summer
Sun Required:	Full sun
Propagation:	Bulblets or bulbs
Plant Type:	Perennial

LEAVES:

The leaves are gathered spring and fall. In the South, wild onion and garlic will continue to sprout out leaves during the winter. They may be used as domestic onions, for seasoning or raw in salads.

Outdoor Tip: The *Army Survival Manual* recommends using the crushed leaves of garlic to relieve the discomfort and itching caused by insect bites.

UNDERGROUND BULBS:

The bulbs are found all year. Usually they are gathered in the second year when they are large enough to use like cultivated onions. The bulbs are very strong, so you do not need much. Bulbs can be used raw, boiled, creamed, pickled, or for seasoning. Their strong taste can be reduced by parboiling and discarding the water. To freeze onions or garlic, one should coarsely chop, blanch two minutes, drain, pat dry, and place them into plastic bags. The bulbs can also be dried for use as seasoning.

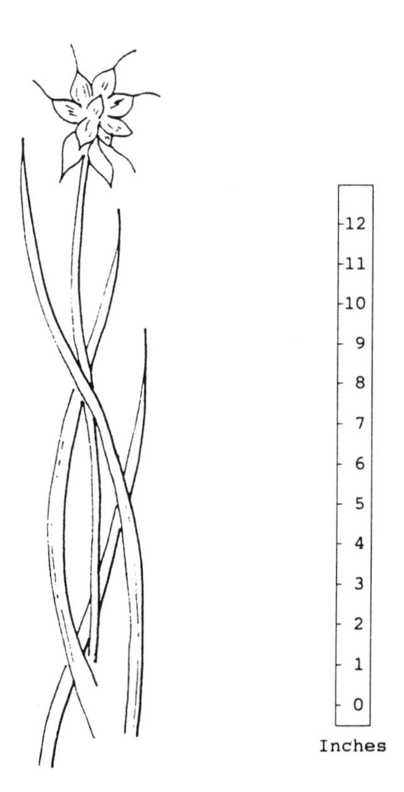

Wild Garlic

Outdoor Tip: Garlic can be eaten to repel insects and chiggers, or the raw bulb rubbed on exposed skin as an insect repellent.

FLOWER STEM BULBLETS:

The bulblets are collected during the summer to flavor soup or for pickling. They also can be sprouted in the same manner as seeds.

Peppergrass

IDENTIFICATION:

Peppergrass is a member of the mustard family with peppery-tasting seeds. Its leaves can vary from area to area. The seeds are flatly rounded and radiate out from a singular stem. Cow cress and poorman's pepper are considered here.

Cow Cress–*L campestre*–Cow cress has elongated, oval leaves that form a rosette at the base of the seed stalk or stem. The leaves become more lance-shaped higher up on the stalk and clasp the stalk with no leaf stem of their own.

Poorman's Pepper–*L virginicum*–Poorman's pepper leaves are mostly deeply toothed, and the stalk leaves have stems.

Peppergrass

Lepidium spp.

Both species have flower or seed stems that range from 6 inches in poor soils to 36 inches in better growing conditions.

RANGE AND ENVIRONMENT:

Peppergrass is found throughout the country but is more rare in the Northern United States. It prefers dry disturbed soils and will grow in harsh conditions.

IDENTIFICATION BRIEFS:

Species:	*Lepidium spp.*
Other Names:	Cow Cress and Poorman's Pepper
Plant Size:	6-36 inches tall
Flower Color:	White
Blooms:	Summer to fall
Sun Required:	Full sun
Propagation:	Seed
Plant Type:	Annual

LEAVES:

The young, sparse, pungent leaves are gathered during early spring and added to salads or cooked like greens. Their use can add "authority" to milder greens. Young leaves should be picked before seeds appear for less bitterness.

SEEDS:

The green seeds are collected summer to fall and are added to season salads, other greens, or soup. They can be removed from the stems by rubbing between the hands. The pods and seeds are normally used together because of the difficulty in separating them. Peppergrass seeds can be stored after drying, or seeds can be ground after drying for table seasoning.

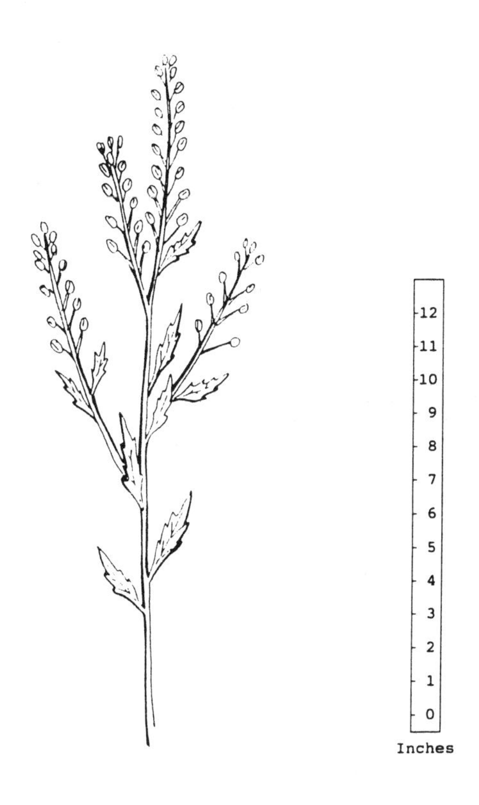

Peppergrass

Pine Tree

IDENTIFICATION:

Pines trees are tall (50-70 feet) with a rough, scaly bark. Needles are evergreen and grow in clusters of 2-5. Pine sap is sticky and has a turpentine smell. All pines are edible, but some species, such as white pine, are not as resinous tasting as others. In addition, the spruce tree can be used like pine trees.

RANGE AND ENVIRONMENT:

The pine tree is found widely in the United States while Scots pine is found in Europe. Both prefer acid soils and rich woods.

IDENTIFICATION BRIEFS:

Species:	*Pinus spp.*
Plant Size:	50-70 feet tall
Flower Color:	Not applicable
Blooms:	Not applicable
Sun Required:	Full sun or semi-shade
Propagation:	Seed
Plant Type:	Perennial

Outdoor Tip: Pine knots (pitch) can be found in old stumps. Break off dead dry limbs right next to the tree trunk to find the pitch wood. Pitch wood is excellent for fire starting and is loaded with turpentine. The resin can be used to waterproof articles. It can also be used as glue. If there is not enough on the tree, cut a V notch in

Pine Tree

Pinus spp.

the bark so more sap will seep out. Put the resin in a container and heat it. The hot resin is your glue. Use as is, or add a small amount of ash dust to strengthen it. Use immediately.

Note: Long leaf pine (*Pinus palustris Mill*) is considered by some to be potentially toxic so should be used in the beginning with caution. Most sources, however, make no distinction between the various pines.

Survival Tip: For a quick meal, peel off the bark of thin twigs. The juicy, strong tasting inner bark can be chewed; it is rich in sugar and vitamins, especially in the spring when the sap is rising. For food, the tender, resinous-tasting new shoots (with needles removed) and the flower clusters can be boiled. Simmering in sugar water improves the strong taste. Young pine seedling roots are gathered during spring and summer. To eat, boil in several changes of water for 30 minutes each.

NEEDLES:

The pine needles maybe gathered all year for tea. Chopped into small pieces and added to boiling water, they should be allowed to steep for five minutes, then the liquid strained. The early spring needles are less resinous. The slightly resinous taste may be masked with a sweetener or in a soup. Pine needle tea has a vitamin C content many times that of orange juice plus it is available all year fresh from the tree.

For a white pine tea, pour 1 cup boiling water over finely chopped needles, covering the top of your cup so the volatile vitamin C will not escape. Add sweetener if desired, let steep for 5 minutes, strain out the needles and enjoy.

Survival Tip: Raw pine needles can be chewed for vitamin C. The brown tips of green pine needles can be used to make a crude survival coffee; the brown tips are broken off, boiled, strained, and then used. Also, cedar needles can be crushed and used to repel insects.

Outdoor Tip: Hardened pine rosin (white pine is best) can be chewed as a mouth freshener.

POLLEN/SHOOTS/FLOWER CLUSTERS:

The pollen, shoots, and flower clusters are gathered during the spring. The high-protein pollen can be shaken out of the male anthers (male cones) and eaten cooked, or added to wheat flour as a protein booster. It may also be sprinkled on other foods or cereals.

Survival Tip: The male pollen anthers (male cones) can be eaten raw and are high in protein but are strongly resinous. They are better if boiled until they sink and then mixed with something else. Salt for a more palatable flavor or bake until they become crisp and brown.

NUTS (PIÑON PINE):

The Western piñon pine nuts are collected in the fall. The large, easily gathered, edible nuts are hidden in the pine cones.

NUTS (EASTERN/SOUTHERN PINE):

The nuts of Eastern and Southern pines are collected in the fall. They have small winged seeds (one-eight inch in diameter) enclosed in a hard protective shell.

When the cones open, the seeds will fall out. If closed, the cone can be heated, causing it to open so the nuts can be shaken out. If one is desperate, the wings can be removed and the very small nut and its coating can be ground together to make a flour for use in a soup or another food.

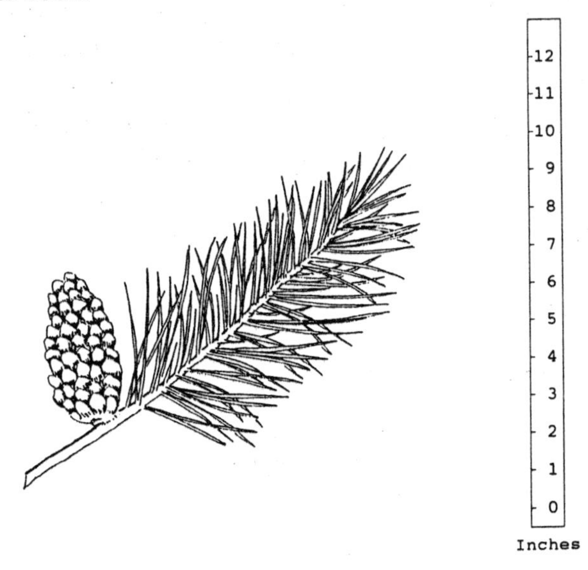

Pine Tree

Survival Tip: Young pine cones can be crushed and used to flavor game and stews. Centers of young green cones can be roasted, making a syrupy food. Pine seeds can be sprouted for additional nutritious food.

PINE BARK

Outdoor Tip: The dry, mature scaling bark on the lower part of a pine cone base makes good dry fire

Broadleaf Plantain

Plantago spp.

tinder. Try to use only cones on the tree because they are drier than those on the ground.

Survival Tip: The inner cambium layer of pines is not always pleasant-tasting with its piney flavor, but it is nutritious. White pine seems to have the best taste. The younger trees seem to be more resinous and stronger tasting. In Europe, during war times, a "famine bread" was made from pine inner bark.The inner bark of pines can be used all year. Pine and spruce have an inner cambium layer (between the outer bark and the inner wood) that can be stripped out and eaten raw or dried outside or smoked by a fire and ground into flour. This flour can be used to extend other flour on a 1:1 basis or as a soup thickener. The cambium layer can also be cut into strips and used as a spaghetti substitute. Another use is to crush the cambium layer pulp, mix it with water, and make it into cakes which are baked for an hour at low temperature. After baking, the cakes may be smoked over the fire to add flavor, then soaked in water to be softened for eating. Or Small cakes can be made from the soft cambium or "new growth" part of a branch or trunk. Crush it and form with water. The "ash cakes" are baked in the hot ashes of the fire or on stones.

Plantain

IDENTIFICATION:

Plantain's seed head looks like a miniature, tightly formed, elongated cattail at the end of a thin stem that has no leaves. Plantain will progressively bloom upward to the top of the flower head. After the flowers mature, the green, tightly formed seed head turns brown.

Broadleaf Plantain (*Plantago major*)–The leaves are 8-10 inches long, oval-shaped, grow out of a central base, and have pointed tips. The leaves have pinnate veins (predominant veins running parallel with the primary stem), and are attached to a short, concave stem. The seed head grows three inches long.

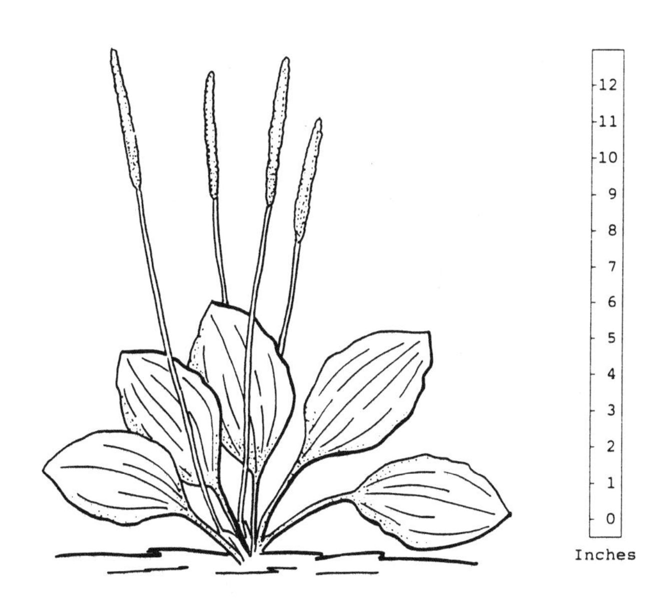

Broadleaf Plantain

English Plantain (*Plantago lanceolata*)–The English plantain or narrow-leaf plantain has lance-shaped leaves. The seed head grows one inch long. English plantain was brought to the United States as a garden vegetable by the colonists.

Plantains are one of my favorite plants. There is something fascinating about a plant brought first to this country as a garden vegetable by our ancestors! Where I live, plantains grow most of the year and are easy to collect. Near my barn the rich soil produces several giant broadleaf plantains! Each year I enjoy watching their white blooms turn into seeds. K.L.

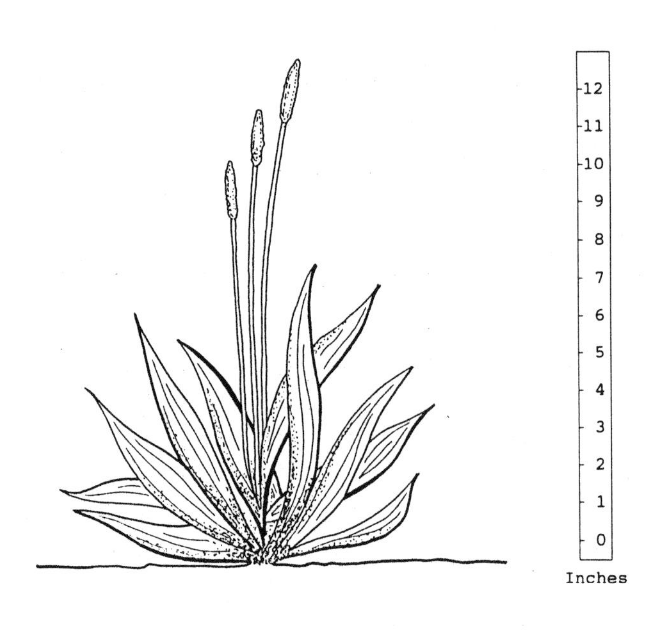

English Plantain

RANGE AND ENVIRONMENT:

Plantain is found throughout in disturbed, poor, moist soils. English plantain winters well in the South and is normally more tender that broadleaf plantain.

IDENTIFICATION BRIEFS:

Species:	*Plantago spp.*
Other Names:	Common Plantain, Ribwort Plantain, Hen-plant, and Soldier's Herb
Plant Size:	8-10 inches tall
Flower Color:	White
Blooms:	Spring
Sun Required:	Full sun or semi-shade
Propagation:	Seed
Plant Type:	Perennial

LEAVES:

The tender, young leaves are gathered early spring through summer and make excellent mild-flavored greens. As summer approaches, the predominant veins in the leaves become stringy, and the plantains require more cooking. They may need to be de-veined or can be finely chopped or blended.

Use to tone down stronger-flavored greens like wild mustard. Plantain leaves can be eaten fresh or cooked as greens. They are best before a flower stalk appears.

To make a tea from fresh or dried leaves, use 1/4-1/2 cup of green leaves and add to a cup of boiling water. Allow to steep for 1/2 hour, strain and use. The leaves may be dried, ground, and also used as flour.

Plantain leaves are rich in vitamins A and C, beta-carotene, and supply a number of required minerals for your diet.

SEEDS:

The seeds are collected spring to summer. They are tiny–smaller than sesame seeds. They can be parched, dried, ground into flour, and used as a soup thickener. They can be dried, boiled, and made into a hot cereal or mush. Plantain seeds are high in B vitamins and vitamin E.

Outdoor Tip: Several seed heads eaten daily can act as a natural insect repellent.

Survival Tip: Native Americans have carried plantain seeds as a survival food when traveling on long hunts.

Pokeweed

IDENTIFICATION:

Pokeweed is a large, perennial native plant that grows up to eight feet tall. The leaves are large, alternate, with large stems. When the plant ages, the stems and leaf veins become red tinged. At maturity, the drooping white flowers turn into shiny purple or black berries.

Warning: Berries, roots, and mature plants are considered poisonous, therefore, leaves are best used as new, young growth. Also, any red-tinged plant material should be discarded. To avoid possibly collecting part of the toxic root, do not cut below ground level.

Pokeweed

Phytolacca americana

As I mentioned before, pokeweed was one the first plants I found, and I like pokeweed! It's interesting to talk to people all over the country and find them in two camps: those who were poor and had pokeweed as a primary vegetable now do not like it, but the others seem to look forward to gathering pokeweed each spring. When I was assigned to Hurricane Hugo near Charleston, S.C., I passed along the edge of a wooded area where I saw a woman and five children picking pokeweed. It must have been a favored green, for this woman had four bushels! It seems to be a neighborhood event there in the spring. When they pick pokeweed, the kids enjoy seeing who can find and collect the most greens. K.L.

RANGE AND ENVIRONMENT:

Pokeweed is found throughout the United States in disturbed soils, often along roadsides and at the edge of woods.

IDENTIFICATION BRIEFS:

Species:	*Phytolacca americana*
Other Names:	Inkberry, Poke Sallet, Pigeonberry, and Pokeberry
Plant Size:	6-8 feet tall
Flower Color:	Greenish, white
Blooms:	Summer
Sun Required:	Full sun or semi-shade
Propagation:	Seed or division
Plant Type:	Perennial

SHOOTS:

The shoots are found during the spring around the old pokeweed stalks. Tender young shoots less than eight inches tall can be peeled, parboiled in two changes of water several minutes each, boiled in a third water until tender, and served like asparagus. The poisonous roots can be gathered after frost, cut into six inch lengths, and planted in sandy dirt in a heated space. They will provide shoots for several months.

STALKS:

The stalks are used during the spring. Young stalks less than one foot tall, with leaves removed, and before red tinged, can be cut and rolled in corn meal and fried like okra. They can also be pickled.

LEAVES:

Young leaves are gathered during the spring, taken from stalks less than one foot tall. When parboiled in two changes of water for several minutes each and boiled in a third water until tender, are especially good with added butter, vinegar, and bacon bits. Another favorite method is to add slices of hard boiled eggs or to scramble several eggs with the greens and sprinkle with vinegar. A large quantity should be gathered because pokeweed does lose bulk during cooking. To freeze, parboil leaves twice, cool, pat dry, and place them in plastic bags. Very high in vitamin C.

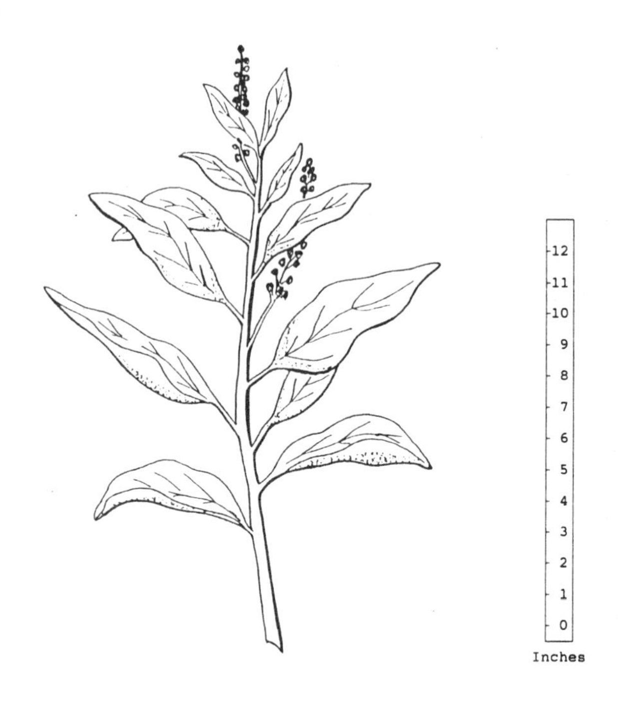

Pokeweed

Prickly Pear

IDENTIFICATION:

Prickly pear is a jointed, desert-type cactus with clumps of flattened leaf pads having bumps of tiny, barbed bristles. Some species have very pointed thorns.

Prickly Pear

Opuntia humifusa

The flowers are yellow followed by a dull red fig-type fruit sometimes called Indian fig.

Prickly pear cactus is commonly served throughout the Southwestern United States. The Wall Street Journal once proclaimed the fruit as the "kiwi fruit of the 1990's." Most is imported from Mexico and South America. The prepared products such as jellies sell the best. The fresh product on store shelves does not sell well due to uninformed customers.

RANGE AND ENVIRONMENT:

Prickly pear is found mostly in the Southern United States in dry sandy soil, but when transplanted survives in most soils.

IDENTIFICATION BRIEFS:

Species:	*Opuntia humifusa*
Plant Size:	3-5 feet tall
Flower Color:	Yellow
Blooms:	Summer
Sun Required:	Full sun
Propagation:	Seed or division
Plant Type:	Perennial

LEAF PADS:

The tender young leaf pads are used during the spring. They can be peeled and the pulp cut into chunks or strips and cooked like string beans. Also, the pad can be battered, roasted, or fried. The interior of the pad is similar to okra and can be used to thicken soups. The pads can be cut into pieces and used raw in salads as a vitamin C source or sliced thin (like green beans) and boiled. The first water is poured off to reduce sliminess. Then once cooking in the second water, you can season with butter and garlic powder. When prickly pear cactus is cooked, it turns dull green. At that time, try adding diced onions and eggs for an unusual dish.

The bristles should be removed before use with a flame or by wiping off with a glove or damp cloth. Otherwise, bristles can be removed by baking the pads in a medium-temperature oven for one-half hour, then peeling the skin with the bristles attached. If a knife is used to cut out the bristles, wipe after each cut, because

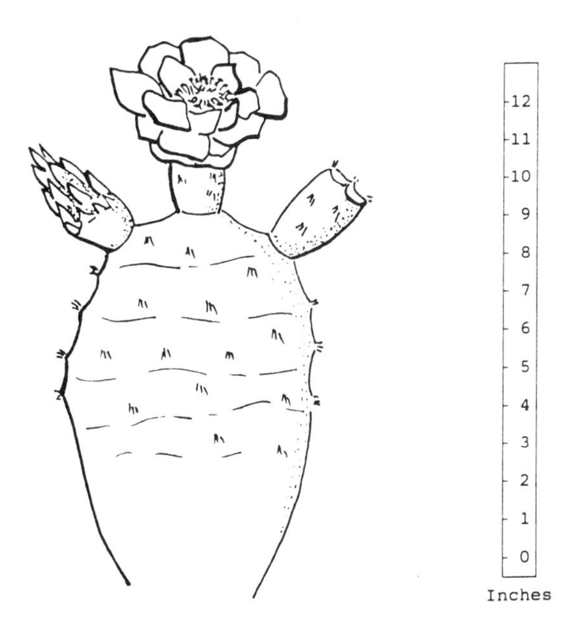

Prickly Pear

of the mucilage produced, the spines will stick to the blade.

The pads are easy to grow. Just plant and they will root.

Outdoor Tip: When camping, roast the pads in their skin on a fire for about 15-20 minutes per side. When cooked, peel, and eat the pulp.

Survival Tip: The pads are 90% water. The easiest way to obtain the water is to simply peal the pads and eat.

FRUIT:

The fruit is collected during summer through fall, and is reddish-purple when ripe. It is best when at this color or even when shriveled up a bit. The fruit should be peeled or cut in half and the pulp scooped out before use. It can be chilled to be eaten raw or pickled after seeds are removed.

SEEDS:

The seeds are collected from the fruit early summer through fall and used as grain. The dried seeds can be crushed or ground into flour and used in soup as a thickener.

Purslane

Portulaca oleracea

Purslane

IDENTIFICATION:

Purslane is a low, prostrate, trailing plant 2-5 inches tall similar in nature to chickweed. The stems frequently fork and grow 12 inches long. The leaves are smooth, ovate (oval), paddle-shaped, thick, juicy, and are 1/2-2 inches long. They grow attached to stems with almost no stalk between the stem and the leaf. The stem sometimes has a reddish tint, and its thickness depends on growing conditions.

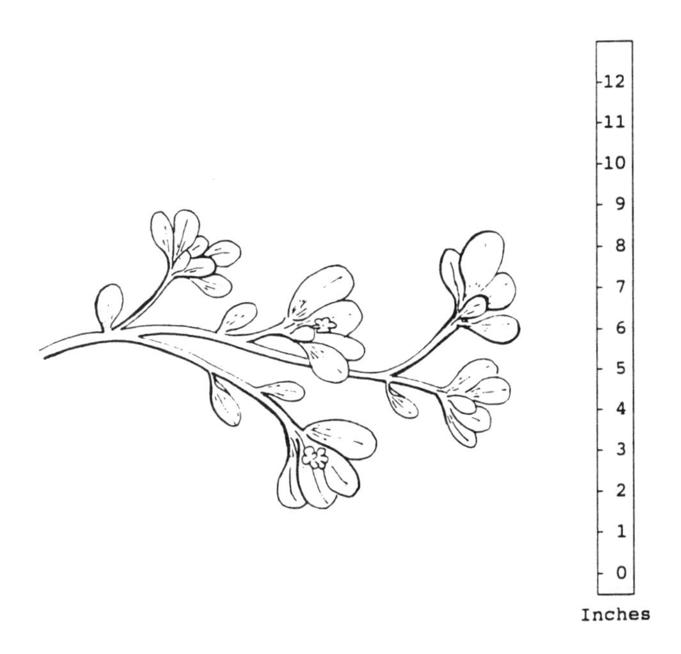

Purslane

The flowers are tiny, five-petaled, yellow, and they grow out of forks where the stems branch. Purslane is one of the better wild edibles. During colonial times, purslane was domestically grown.

Purslane has Omega-3 fatty acids that help thin the blood, lower serum cholesterol, and protect the heart.

Several years ago while on a flooding assignment in Miami Springs, Florida, I found a patch of purslane growing along the edge of the sidewalk. I do not see much purslane around my home, so each day on my way to work, I checked out the plant's progress. Later after Hurricane Andrew hit south of Miami, I was located only two miles from where I had been before. I went over to the same sidewalk and found the same plant as healthy as two years before. Once you find a concentration of wild edibles, they will provide for you for many years. K.L.

RANGE AND ENVIRONMENT:

Purslane is found throughout in light, fertile, sandy disturbed soils located in waste places.

IDENTIFICATION BRIEFS:

Species:	*Portulaca oleracea*
Other Names:	Pursley, Pusley, and Pigweed
Plant Size:	2-5 inches tall
Flower Color:	Yellow
Blooms:	Summer to fall
Sun Required:	Full sun or semi-shade
Propagation:	Seed
Plant Type:	Annual

LEAVES AND STEMS:

The leaves are used during all the summer. The succulent, okra-like mucilaginous leaves are added to salads or boiled in salt water as a vegetable. They can be sautéed with meat, onions, or corn. Very little of purslane's bulk is lost during boiling. If only a small concentration of purslane is found, snip off the leafy tips, allowing the plant to continue to produce new leaves until early fall. Purslane is good fresh, steamed, frozen, and pickled. To freeze, blanch for two minutes, drain, pat dry, and place in plastic bags. Try growing it during the winter in a greenhouse or use the dried leaves like a pot herb.

Purslane is a good source of beneficial omega-3s fatty acids normally found in oily fish. In addition, it has more vitamin A than any other vegetable.

SEEDS:

The seeds are collected during late summer. With patience, the large production of minute, black seeds can be ground into flour or cooked into gruel. To gather the seeds, place plants on a large cloth and dry for several weeks. Next crush the plants and use a strainer or window screen to separate the seeds. Or, just dry the plant and shake it over a large piece of paper to collect the seeds.

Wild Rose

Rosa spp.

Wild Rose

IDENTIFICATION

The leaves are finely toothed and spaced alternately on a thorny stem that grows 3-15 feet in height. The rose hips themselves are reddish-orange in color when ripe and range in size from that of a pea to the size of a nickel. Different varieties of rose hips differ in flavor, and wild roses produce smaller rose hips than domestic ones.

RANGE AND ENVIRONMENT:

Roses are found throughout in open woods, pastures, old homesteads, and moist soils along streams.

IDENTIFICATION BRIEFS:

Species:	*Rosa spp.*
Plant Size:	3-15 feet tall
Flower Color:	Various
Blooms:	Summer
Sun Required:	Full sun
Propagation:	Seed
Plant Type:	Perennial

PETALS:

The petals are found during the summer and are eaten raw or added to salads or other foods for a special flavor. For a hot tea rich in vitamin C, use two teaspoons of fresh petals or one teaspoon of dried ones. The petals can be dried for future use, but the bitter flower base must be removed before use.

Outdoor Tip: Rose petals can be applied over a wound to make a field dressing. As they dry, they form a protective scab.

FRUIT:

Rose hips are collected fall through early winter when fully colored. It is best to pick them after frost. The vitamin C and sugar contents of fresh or dried rose hips are often more concentrated than that of orange juice. The rind of rose hips can be eaten raw or softened by boiling, or the whole hip can be steeped to make a tea. Use one rounded teaspoon of crushed dried hips for a flavorful tea. Hairs that surround the seeds can irritate intestinal linings unless the tea is strained.

To store, the rose hips should be dried very slowly. After drying, they can be stored whole or ground. Always test for dryness by placing the hips in a sealed jar in the sun for several hours. If moisture collects on the jar's wall, then additional drying is needed, or the hips will mold.

It was by accident that I found my first wild rose hips. I had been looking for them for several months, but all I had found were wild roses with no fruit. One evening while in Birmingham, Alabama, on a flooding assignment, I was taking an early evening walk in an industrial park. There along the parking lot in an overgrown lot was the largest wild rose I have seen. It was full of rose hips. There must have been hundreds. All

I can figure is that the heavy people traffic had kept the birds away. All wild foods are not in the woods! K.L.

Survival Tip: In Europe, military survival schools taught pilots to carry rose hips in their pockets and eat them as a stimulant when under heavy physical strain or stress.

LEAVES:

The leaves are collected spring through summer and can be steeped to make a tea.

Wild Rose

Sassafras

Sassafras albidum

SHOOTS AND SEEDS:

The shoots are gathered during the spring and can be eaten raw or cooked like a vegetable. The seeds are collected during the winter. The dried inner core of seeds from a rose hip can be ground into flour or cooked into a mush. The seeds contain vitamins C and E.

ROOTS:

The roots are used all year and can be cleaned and steeped into a tea.

Sassafras

IDENTIFICATION:

Sassafras normally grows into a small tree but can grow to fifty feet tall. The reddish-brown bark is deeply grooved and in older trees gives the appearance of cork. Sassafras is easily identified by its three types of leaves, all of which are toothless. One leaf has two lobes and looks like a winter mitten with one thumb. Another leaf will have three lobes while the third leaf is lance-shaped. All leaf shapes are found on the same tree and when crushed give off the familiar sassafras smell.

The fruit is small, round, and blue/purple on reddish stems. Its vitamin C was once used to cure scurvy.

Warning: There are mixed opinions regarding limited research that has shown sassafras contains a carcinogen which in large quantities has caused cancer in laboratory animals.

RANGE AND ENVIRONMENT:

Sassafras is found in Eastern United States along the edge of woodlands.

IDENTIFICATION BRIEFS:

Species:	*Sassafras albidum*
Plant Size:	Up to 50 feet tall
Flower Color:	Not applicable
Blooms:	Not applicable
Sun Required:	Full sun or semi-shade
Propagation:	Seed or division
Plant Type:	Perennial

LEAVES AND BUDS:

The tender leaves are gathered spring through fall and are eaten raw. The young leaves can be dried slowly in the sun and crushed to make a powder which can be used to thicken and flavor stews and soups or placed in a salt shaker and used like a spice. Store in air-tight jars in a dark place. The buds are collected during the winter and used in salads.

"Gumbo file" which is made of 100% sassafras can be found in the spice area of a grocery store.

Outdoor Tip: To remove any bad odors from your hands, rub and crush the fresh leaves between your hands. A strong tea can be made from the leaves and rubbed on body parts to act as a natural insect repellent. The same tea can be used as a good antiseptic mouthwash and freshener. A custom fitting toothbrush can be made by chewing sassafras twigs.

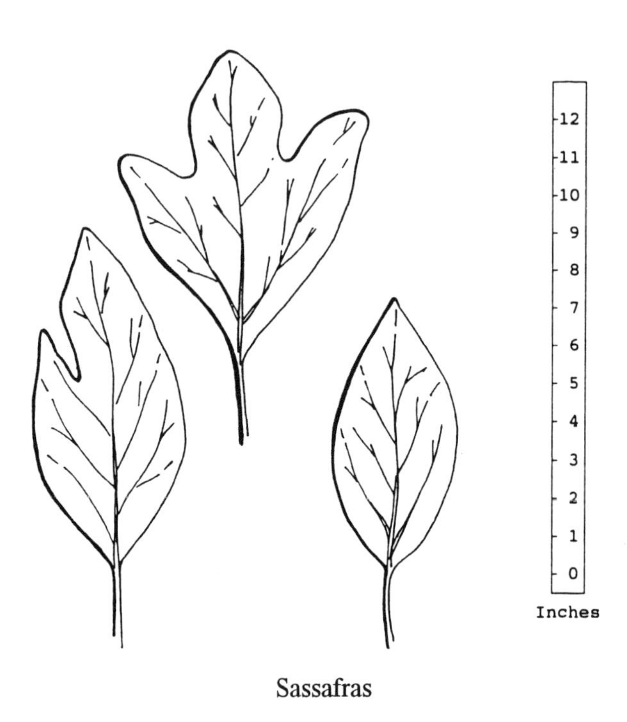

Sassafras

ROOT BARK:

The sassafras root bark is used fall to winter. The roots are well known for the tea they produce which is better than a tea made from the leaves. Roots for tea are collected during colder months from dormant trees because during the growing season they tend to be bitter. Strip off the bark and dry it in an oven at low temperature. Store the bark in an air-tight container. To make a

Shepherd's Purse

Capsella bursa-pastoris

tea, boil the root bark until the water turns reddish-brown; sweeten and drink it hot or cold. The root bark is also used to flavor meats. Some prefer not to bother with preparing root bark and use roots of less than one inch in diameter which can be boiled again for making several batches of tea.

Shepherd's Purse

IDENTIFICATION:

Shepherd's purse leaves grow in a rosette with deeply scalloped leaves similar to those of dandelion. As the leaves travel up the seed stem, they become elongated. The leaves have no stems of their own and directly clasp the plant stem that grows to two feet tall. Shepherd's purse has white flowers that are followed by flat, heart-shaped seeds resembling heart-shaped purses.

RANGE AND ENVIRONMENT:

Shepherd's purse is found throughout along roadsides and in other disturbed soils.

IDENTIFICATION BRIEFS:

Species:	*Capsella bursa-pastoris*
Other Names:	Caseweed
Plant Size:	Up to two feet tall
Flower Color:	White
Blooms:	Summer

Sun Required: Full sun
Propagation: Seed
Plant Type: Annual

LEAVES:

The sparse leaves of shepherd's purse are gathered for best flavor during spring through early summer and before flowers appear. Young leaves can be used in a salad where they tend to shed any dressing used. They may also be used as a cooked green. The leaves are

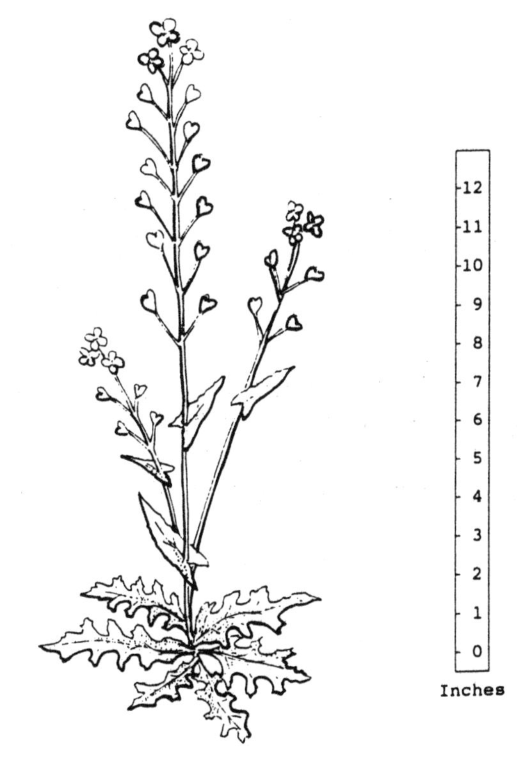

Shepherd's Purse

peppery flavored and may require parboiling with the initial water discarded before final cooking. Leaves can add flavor to other vegetables and stews, and they can be stored for future use as a seasoning.

SEEDS:

The dried seeds are collected during the summer and used as a pepper-like seasoning in stews. The ground seeds can be made into a nutritious meal or mixed with vinegar to make mustard. Dried seeds can be stored for future use.

Sorrel, Sheep

IDENTIFICATION:

Sheep sorrel grows 4-10 inches tall depending on growing conditions. The leaves are bright green, arrow-head-shaped, and have basal lobe leaf extensions. They have a distinctive juicy, sour taste. The green seeds acquire a reddish tint as they mature. When, during early spring to summer, you pass a fallow field that has a reddish tint, you can assume it is full of sheep sorrel that is going to seed. The seed stalk is erect, twelve inches tall with rows of closely spaced reddish flowers and seeds. Early settlers used sheep sorrel in soups, salads, and drinks. The plant is rich in vitamin C.

WARNING: Sorrel contains oxalic acid as does spinach. If eaten in large quantities, sorrel may cause stomach cramps and can inhibit the absorption of calcium

Sheep Sorrel

Rumex acetosella

in the body. A normal serving will do no harm, and cooking dissipates the oxalic acid.

RANGE AND ENVIRONMENT:

Sheep sorrel is found throughout along highways and in fields having disturbed sandy, poor, acid soils. Sheep sorrel is a good indicator of sour soil.

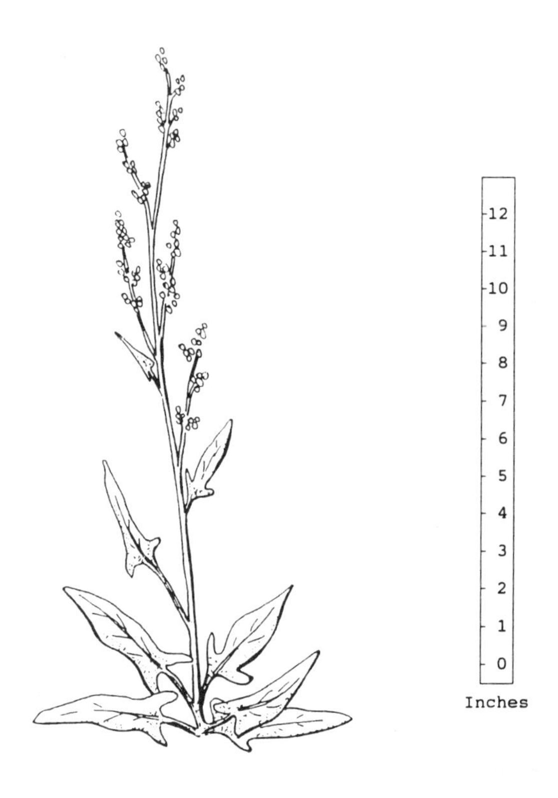

Sheep Sorrel

IDENTIFICATION BRIEFS:

Species:	*Rumex acetosella*
Other Names:	Red Sorrel, Common Sorrel, and Sour Grass
Plant Size:	4-10 inches tall
Flower Color:	Reddish
Blooms:	Spring to summer
Sun Required:	Full sun, semi-shade
Propagation:	Seed
Plant Type:	Perennial

LEAVES:

The leaves are gathered early spring through early summer and are excellent, fresh, in salads because of the pleasant sour taste. The leaves can be added to soup and stews. They can be steeped into a liquid that can be used as a vinegar replacement or can be sweetened to create a lemonade-type drink for gaining vitamin C. Leaves are good for stuffing fish and can be added to curly dock, chickweed or other bland greens to pep them up. As leaves mature, they lose their sourness. To freeze, blanch for one minute, drain, pat dry, and pack in plastic bags.

Outdoor Tip: The process of nibbling on the sour leaves acts as a thirst quencher.

FLOWERS AND SEEDS:

The flowers are used with the leaves spring through summer in salads or cooked. Also, the seeds are collected spring through summer. They are very small but can be ground into meal. Boiling the seeds adds variety and a red color to liquids or soups.

Sorrel, Wood

IDENTIFICATION:

The leaves of wood sorrel are similar in size and appearance to clover. The key for identification is that wood sorrel has a sour taste and a heart-shaped leaf, whereas clover has a rounded leaf. Wood sorrel prefers acid soil. Wood sorrel is high in vitamin C and was used in the past to prevent scurvy. We will discuss two types of wood sorrel.

Violet Wood Sorrel–Violet wood sorrel has leaves that can reach the diameter of a quarter. The leaves tend to close at night and on cloudy days. The flower has five petals and is purple in color.

Yellow Wood Sorrel–The leaves of yellow wood sorrel are smaller than those of violet wood sorrel and not as green. The overall plant is smaller than violet wood sorrel and is more fragile. The five-petaled flower is yellow. The stems are tougher than stems of violet wood sorrel; therefore, only the leaves and flowers should be used.

WARNING: Wood sorrel also contains oxalic acid. If eaten in large quantities, sorrel like spinach, can inhibit the absorption of calcium in the body. A normal serving will do no harm, and cooking helps dissipate the oxalic acid.

Wood Sorrel

Oxalis stricta

RANGE AND ENVIRONMENT:

Wood sorrel is found widely in moist, acid soils in semi-shaded areas.

IDENTIFICATION BRIEFS:

Species:	Yellow Wood Sorrel–*Oxalis stricta*. Violet Wood Sorrel–*O. violacea*. Plus others.
Other Names:	Sour-grass
Plant Size:	3-6 inches tall
Flower Color:	Yellow, violet
Blooms:	Summer to fall
Sun Required:	Semi-shade
Propagation:	Seed
Plant Type:	Annual

LEAVES:

The leaves are gathered early spring through fall. The tender stems and leaves can be steeped in hot water (to save vitamin C, do not boil). Use the liquid as a sour lemonade-type drink or as a vinegar replacement. For a tea, use a handful of leaves per pint of water. Wood sorrel is excellent in salads, adding a lemony taste. Cook with greens to perk up the taste of milder ones. The stems may be removed if too stringy for use.

Outdoor Tip: The process of nibbling on the sour leaves acts as a thirst quencher.

FLOWERS AND SEEDS:

The flowers are collected during the summer and used with leaves in salads and as cooked greens. The seeds pods are collected summer to fall. They follow the flower, are small, and look like a pea pod that is 1/2 to 1 inch long. The young seed pods can be added to salads or cooked with the leaves and stems. The small seed pods, being sour and crunchy, make one of my favorite snacks.

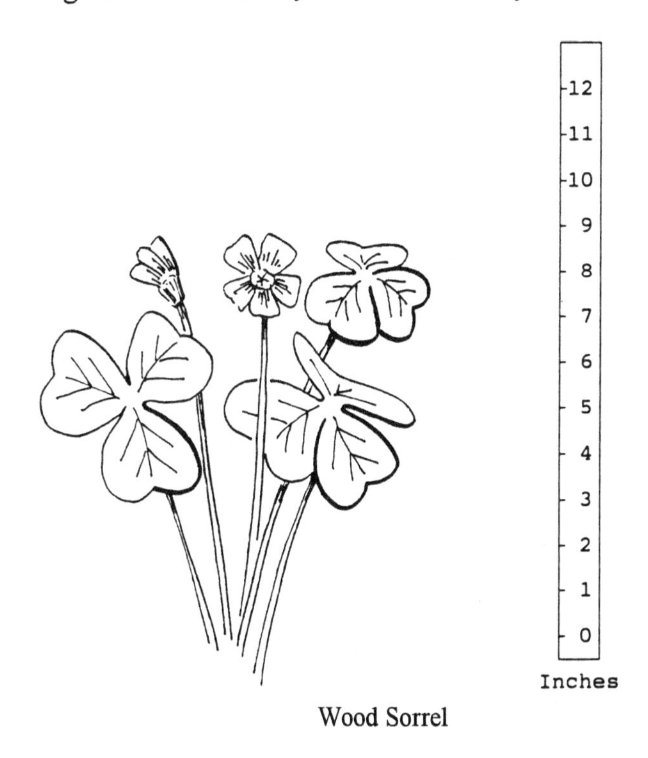

Wood Sorrel

TUBERS/ROOTS:

The tubers and roots are gathered all year. They are cleaned and eaten raw or cooked with the greens, seeds, and flowers.

Wild Strawberry

IDENTIFICATION:

Wild strawberries are similar to domestic strawberries, but the fruits are smaller. Each stem has three leaves which have saw-toothed edges. Flowers are white with five rounded petals.

Similar snakeberries are bumpy with red seeds rather than pot-marked with golden seeds as are wild strawberries. Snakeberries have small, hard, bitter fruit which should not be used.

RANGE AND ENVIRONMENT:

Wild strawberries are found throughout, preferring fertile soil in meadows and open woods.

IDENTIFICATION BRIEFS:

Species:	*Fragaria spp.*
Plant Size:	4-6 inches tall
Flower Color:	White
Blooms:	Spring
Sun Required:	Full sun or semi-shade
Propagation:	Division
Plant Type:	Perennial

FLOWERS AND FRUIT:

The flowers are found during early summer and used raw in salads. The fruit is collected during early summer and used fresh or in jams.

Wild Strawberry

Fragaria spp.

LEAVES:

The young leaves are used during the summer and eaten raw. A tea made from fresh or dried leaves is very high in vitamin C. Use the dried leaves also to make flour.

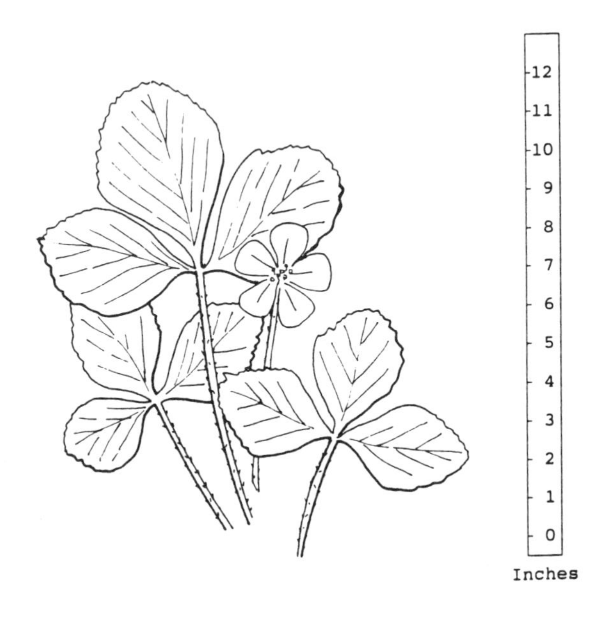

Wild Strawberry

Thistle, Bull

IDENTIFICATION:

Bull thistle is very spiny with deeply cut leaves and yellow-tipped spines. The plant grows to six feet tall and,

Bull Thistle

Cirsium vulagare

depending on the variety, will have mostly straight stalks. The flowerheads are easy to recognize because of the bristle-brush shape and bright red-purple color. After maturing, the flower head turns white and resembles a dense dandelion seed puff. Canada thistle can also be used as a wild edible but its stalk is hollow.

RANGE AND ENVIRONMENT:

Bull thistles are found throughout in meadows, pastures, roadsides, and other disturbed soils. The thistle is commonly found in Scotland and is used as their national emblem.

IDENTIFICATION BRIEFS:

Species:	*Cirsium vulagare*
Other Names:	Spear Thistle
Plant Size:	Up to six feet tall
Flower Color:	Red, purple
Blooms:	Summer
Sun Required:	Full sun
Propagation:	Seed
Plant Type:	Biennial

LEAVES:

The younger leaves are gathered during the spring. Older leaves are unpalatable. First-year plants will not have flower stalks. Leaves from first-year plants are normally the best for eating raw or cooked as greens. The spines must be removed, using gloves and a knife because they are very sharp. The spines can be also burned off by holding the stalk briefly next to a flame. Eating the leaves is an acquired taste.

BUDS:

Both flower buds and flower heads are edible. They are gathered during spring and summer and are eaten like artichokes. For the best results, steam first before eating.

At the base of each flower head is a nutritious "nut" which can be eaten raw.

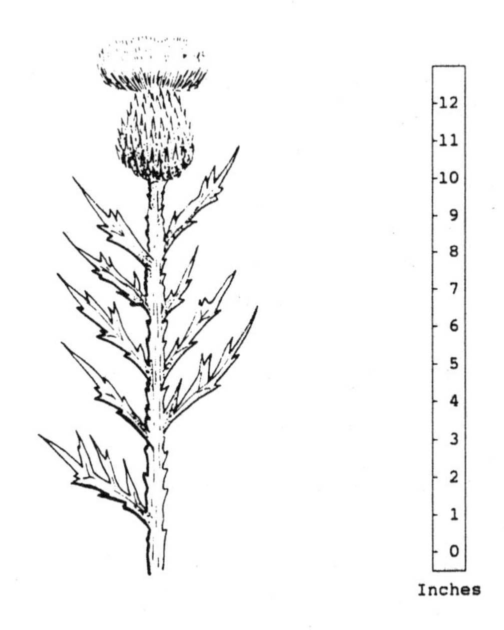

Bull Thistle

Outdoor Tip: The down from dried flower tops makes a good insulation and fire tinder.

STALKS:

The young stalks are used spring to summer, peeled and eaten raw or cooked. The peeled stalks make a refreshing celery-like snack. Thistle stalks add texture to other cooked vegetables or a crunchiness to salads.

Outdoor Tip: The water-rich raw stalks can quench thirst. When the stalk is stringy, it can be chewed to remove the liquid and nutrients and the pulp spit out. Caution: The juice from the thistle is a powerful laxative.

Survival Tip: Excellent cordage made from thistle stalk fibers can be used for fish lines or snares.

ROOTS:

The roots are collected all year. Usually, but not necessarily, the roots are peeled. The roots can be boiled alone, boiled with other foods, fried, or ground into flour. I brush the sliced roots with olive oil, cover with aluminum foil and roast in the oven. Remove the foil during the last few minutes for a crisper texture.

Survival Tip: The roots offer an energy-rich winter survival food.

Sow Thistle

Sonchus oleracea

Thistle, Sow

IDENTIFICATION:

The leaves of sow thistle are sharply toothed and spiny-looking, but they are not thorny. They attach directly to a hairy stalk in a curled or swirl-type clasping fashion that is easy to identify. The leaves and stalk, like wild lettuce, bleed a milky sap. Sow thistle flowers are yellow and resemble dandelion. When the flowers mature, they drop off and are replaced by seeds that resemble the long, white hairs of the dandelion ball of seeds. Sow thistle can grow eight feet tall.

RANGE AND ENVIRONMENT:

Sow thistles are found in disturbed soils of fields..

IDENTIFICATION BRIEFS:

Species:	*Sonchus oleraceus*
Plant Size:	Up to eight feet tall
Flower Color:	Yellow
Blooms:	Summer
Sun Required:	Full sun
Propagation:	Seed
Plant Type:	Annual

LEAVES:

The young leaves are gathered early spring and cooked. Although dandelion is more commonly known, some find that sow thistle is more agreeable, being less bitter and good for energy like spinach. Collect it before blooming for the best greens, but any bitter flavor can be removed by par-boiling, pouring off the water, then cooking the leaves tender. If needed, remove spines with scissors. Like dandelion, best part of salad is crown. To freeze: parboil, drain, pat dry and place into plastic bags.

Violet, Wild

IDENTIFICATION:

Low-growing plants 3-6 inches tall, violets have dark green, long-stemmed, heart-shaped, and smooth-to-

Wild Violet

Viola spp.

lightly-toothed leaves. The solitary flower is five petaled and may be bearded.

Collect violets when in bloom to avoid confusion with similar inedible plants. The blue-flowered violet is commonly the species used due to availability and ease of recognition when blooming. Avoid the violet roots, for they are very emetic. African violets are not related to wild violets and are not edible.

RANGE AND ENVIRONMENT:

Violets are found throughout, preferring moist meadows and semi-shaded areas.

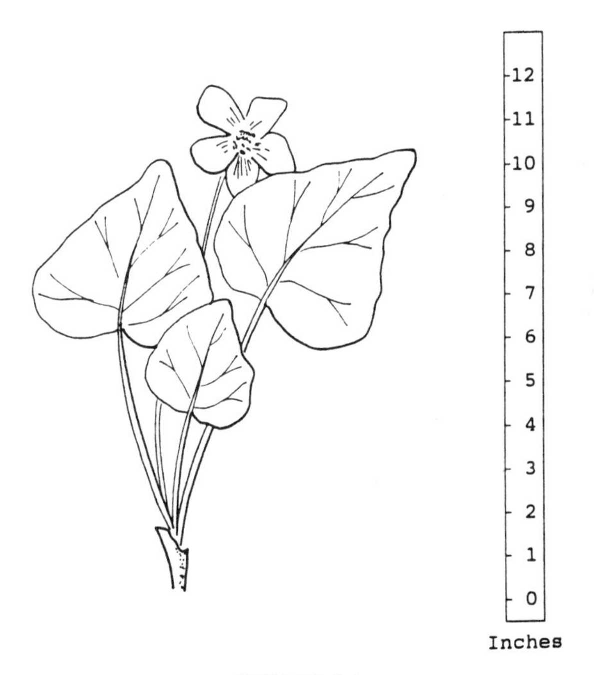

Wild Violet

IDENTIFICATION BRIEFS:

Species:	*Viola spp.*
Other Names:	Wild Okra
Plant Size:	3-6 inches tall
Flower Color:	Blue and yellow
Blooms:	Spring to Summer
Sun Required:	Semi-shade
Propagation:	Seed, division
Plant Type:	Perennial

LEAVES:

The young leaves are gathered in the spring and are excellent raw or cooked in soup or greens. Cook the leaves like spinach and serve with normal seasonings or with vinegar and hard-boiled eggs. Be sure to drink the nutritious liquid. Leaves are mucilaginous, like okra, and used for thickening soup. Dried leaves can be saved and used to make tea. After late fall, the leaves become bitter. Violet leaves are very high in vitamin A, vitamin C, iron rutin, and beta-carotene.

FLOWERS:

The flowers are sweet and mild. They are collected spring through summer and are used fresh in salads or cooked in soup or greens.

CONCLUSION

I hope that you have enjoyed learning about wild edibles and their many possibilities for use. As one studies plants, that nondescript, perhaps annoying "weed" in the field–or lawn–seems to develop special characteristics. First, one will notice identifying distinctions such as leaf shape and texture, smooth or hairy stems, and the color of its flowers. Then the newly sensitized eyes will begin to see other traits of fascination and beauty never before imagined. Also, I hope those readers who know of other uses of wild plants will feel free to pass their information on to me for future publications. Recipes for wild edibles would be especially helpful. K.L.

APPENDIX A

FLOWER COLOR GUIDE

Flower color is used for initial grouping to narrow down the identification process.

Many edible plants have distinctive blooms. In this section, they are grouped by flower color.

To use this guide, first determine the color of the bloom of the plant in question. Then look it up in the matching color category.

WHITE

Arrowhead
Bittercress
Brambles
Carrot, Wild
Prickly Pear
Purslane
Clover
Onion/Garlic
Peppergrass
Plantain
Pokeweed
Shepherd's Purse
Strawberry, Wild

YELLOW

Dandelion
Evening Primrose
Lettuce, Wild
Mustard, Wild
Chickweed, Common
Chickweed, Mouse-ear
Sorrel, Wood
Thistle, Sow

PINK/RED

Burdock
Clover, Crimson
Henbit
Milkweed
Onion/Garlic
Roses
Sorrel, Sheep
Thistle, Bull

VIOLET/BLUE

Chicory
Kudzu
Sorrel, Wood
Thistle, Bull
Violet

GREEN

Cattail
Dock, Curly
Grape
Greenbrier
Lamb's-quarters
Pokeweed

APPENDIX B

SEASONAL GUIDE

Taking advantage of seasonal plant activity, one can plan a profitable foraging trip to collect selected plant parts.

Certain plants can be found most of the year, but when they are not in their prime, they may be bitter or tough. This seasonal guide will list only the prime season for the selected plant parts.

OUTDOOR NOTE: When the seasons for some plants are extended past their prime period, their taste and tenderness may be sacrificed but the nutrients will still be available.

To use the seasonal guide, determine your season, then use the guide to select your plant for the desired parts.

SPRING

Amaranth	Leaves
Bittercress	Leaves
Brambles	Leaves, shoots
Burdock	Leaves, stalks
Carrot, Wild	Leaves, roots, shoots
Cattail	Shoots, roots, flower spikes (late spring), pollen (early spring)
Chickweed	Leaves
Chicory	Leaves, roots
Clover	Leaves, flowers

Dandelion	Leaves (early spring), crowns (early spring), flower buds (early spring), roots (early spring)
Dock, Curly	Leaves (early spring)
Evening Primrose	Leaves, roots
Grapes	Leaves, shoots
Greenbrier	Leaves, shoots, tendrils, roots
Grass	Leaves
Henbit	Leaves
Kudzu	Leaves, shoots, roots
Lettuce	Leaves (early spring)
Lamb's-quarters	Leaves (late spring)
Maple Tree	Seedlings (early spring), inner bark
Milkweed	Leaves, shoots
Mustard, Wild	Leaves (early spring)
Onion/Garlic	Leaves, bulbs
Peppergrass	Leaves (early spring)
Pine	Needles, pollen/shoots/flower clusters, inner bark
Plantain	Leaves, seeds
Pokeweed	Leaves (early spring), shoots (early spring), stalk (early spring)
Prickly Pear	Leaf pads
Roses	Leaves, roots, shoots
Sassafras	Leaves
Shepherd's Purse	Leaves
Sorrel, Sheep	Leaves (early spring)
Sorrel, Wood	Leaves, tubers/roots
Thistle, Bull	Leaves, stalk, roots
Thistle, Sow	Leaves

SUMMER

Amaranth	Seeds
Brambles	Leaves, berries

Burdock	Roots, stalks, seeds (late summer)
Carrot, Wild	Flowers, seeds (late summer)
Cattail	Roots, sprouts/corms (late summer), seeds
Clover	Leaves, flowers
Dandelion	Flower buds
Dock, Curly	Leaves, seeds (midsummer)
Goldenrod	Leaves
Grapes	Leaves, tendrils (early summer), shoots, fruit (late summer)
Grass	Leaves, seeds
Greenbrier	Leaves, shoots, tendrils, roots
Kudzu	Leaves, flowers, shoots, roots
Lamb's-quarters	Leaves, seeds (late summer)
Lettuce	Leaves, flower heads
Milkweed	Flower buds, flowers, seed pods
Mustard, Wild	Flower buds, flowers, seeds, seed pods
Onion/Garlic	Bulbs, bulblets
Peppergrass	Seeds
Pine	Needles, inner bark
Plantain	Leaves, seeds
Prickly Pear	Fruit, seeds
Purslane	Leaves, stems, seeds (late summer)
Roses	Leaves, fruit, petals, roots
Sassafras	Leaves
Shepherd's Purse	Leaves (early summer), seeds
Sorrel, Sheep	Leaves (early summer), flowers
Sorrel, Wood	Leaves, tubers/roots, flowers
Strawberry, Wild	Leaves, fruit (early summer)
Thistle, Bull	Stalk, roots

FALL

Amaranth	Seeds

Arrowhead	Tubers
Bittercress	Leaves
Burdock	Roots
Carrot, Wild	Roots
Cattail	Roots, sprouts/corns, tubers
Chicory	Roots
Chickweed	Leaves
Clover	Roots
Dandelion	Roots
Dock, Curly	Seeds
Goldenrod	Leaves
Grapes	Fruit
Grasses	Leaves, seeds
Greenbrier	Roots
Kudzu	Leaves, shoots, roots
Lamb's-quarters	Seeds
Maple Tree	Seeds
Mustard, Wild	Seeds , seed pods, flower buds, flowers
Oak	Nuts
Onion/Garlic	Bulbs, leaves
Peppergrass	Seeds
Pine	Nuts, needles, inner bark, cones
Plantain	Leaves
Prickly Pear	Fruit, seeds
Roses	Fruit, roots
Sassafras	Leaves, root bark
Sorrel, Wood	Leaves, seeds, tubers/roots
Thistle, Bull	Roots

WINTER

Arrowhead	Tubers
Bitter Cress	Leaves

Cattail	Roots, tubers
Chickweed	Leaves (South)
Chicory	Roots
Clover	Roots
Dandelion	Roots
Grasses	Leaves (dried)
Greenbrier	Roots
Henbit	Leaves (late winter)
Kudzu	Roots
Lamb's-quarters	Seeds (early winter)
Maple Tree	Sap
Mustard, Wild	Seeds, seed pods
Onion/Garlic	Bulbs
Pine	Needles, inner bark
Roses	Fruit, seeds, roots
Sassafras	Buds, root bark
Sorrel, Wood	Tubers/roots
Thistle, Bull	Roots

APPENDIX C
COMPOSITION OF FOODS

	Source	Energy (kcal)	Prot (g)	Fats (g)	Carb (g)	Calc (mg)
Amaranth, raw	1	26	3.7	.3	4	215
Apple	6	59	.2	.4	15	7
Arrowhead	**1**	**413**	**5.0**	**.3**	**20**	**10**
Beans, snap, raw	1	31	1.8	.1	7	37
Beans, lima, fresh,boiled	1	123	7.0	.3	24	32
Beef, chuck, blade 1/4 fat trim	4	237	31.0	11.6	0	13
Beef, liver, pan fried	4	217	26.7	8.0	8	11
Bittercress (Watercress)	**1**	**11**	**2.3**	**.1**	**1**	**120**
Blackberry, fruit, raw	**6**	**52**	**.7**	**.4**	**13**	**32**
Bread, wheat, white	8	361	12.0	1.6	72	15
Broccoli, raw	1	28	3.0	.4	5	48
Burdock, root, cooked	**1**	**88**	**2.1**	**.1**	**21**	**49**
Cabbage, raw	1	24	1.2	.2	5	47
Carrots, raw	1	43	1.3	.2	10	27
Chicory, greens, raw	**1**	**23**	**1.7**	**.3**	**5**	**100**
Chicory, root, raw	**1**	**73**	**1.4**	**.2**	**17**	**41**
Chicken, white, roasted	7	239	27.3	13.6	0	15
Chestnut, Chinese, roasted	5	239	4.5	1.2	52	19
Collards, boiled	1	14	1.1	.2	3	78
Corn, fresh, boiled	1	108	3.3	1.3	25	2
Dandelions, raw	**1**	**45**	**2.7**	**.7**	**9**	**187**
Dock, raw	**1**	**22**	**2.0**	**.7**	**3**	**44**
Eggs, hard cooked	9	155	12.6	10.6	1	50
Garlic, raw	**1**	**149**	**6.4**	**.5**	**33**	**181**
Grapes, raw	**6**	**63**	**.6**	**.4**	**17**	**14**
Kale, cooked	1	32	1.9	.4	6	72
Kudzu, roots	**2**	**113**	**2.1**	**.1**	**ND**	**15**
Lamb's Quarters, raw	**1**	**43**	**4.2**	**.8**	**7**	**309**
Lettuce, Romaine, raw	1	16	1.6	.2	2	36
Lettuce Iceberg, raw	1	13	1.0	.2	2	19
Maple Syrup	**10**	**262**	**0**	**.2**	**67**	**67**
Milk, cow, whole	6	61	3.3	3.3	5	119
Milkweed	**2**	**ND**	**.1**	**0.5**	**ND**	**ND**
Mushroom, raw	**1**	**25**	**2.1**	**.4**	**5**	**5**
Mustard Greens	**1**	**15**	**2.3**	**.2**	**2**	**74**
Oak, acorns raw	**5**	**369**	**6.2**	**23.8**	**40**	**41**
Onions, raw	**1**	**34**	**1.2**	**.3**	**7**	**25**
Onions, spring, raw + tops	**1**	**25**	**1.7**	**.1**	**6**	**60**
Orange, Raw	6	47	.9	.1	11	40

Phos (mg)	Iron (mg)	Sod (mg)	Pot (mg)	A (IU)	Thia (mg)	Ribo (mg)	Niac (mg)	Asor (mg)
50	2.3	20	50	2917	.02	.16	.67	43
7	.2	0	115	53	.02	.01	.08	6
174	**2.6**	**22**	**922**	**0**	**.17**	**.07**	**1.70**	**1**
38	1.0	6	209	688	.08	.11	.75	16
130	2.5	17	570	370	.14	.10	1.04	10
335	3.7	71	263	0	.08	.28	2.67	0
461	6.3	106	364	36105	.21	.14	14.4	23
60	**.2**	**41**	**330**	**4700**	**.09**	**.12**	**.20**	**43**
21	**.6**	**0**	**196**	**165**	**.03**	**.04**	**.40**	**21**
97	4.4	2	100	ND	.81	.51	7.55	0
66	1.0	15	325	1542	.07	.12	.63	93
93	**.8**	**4**	**360**	**0**	**.04**	**.06**	**.32**	**0**
23	.6	18	246	126	.05	.03	.30	47
44	.5	35	323	28129	.10	.06	.93	9
47	**.9**	**45**	**420**	**4000**	**.06**	**.10**	**.50**	**24**
61	**.8**	**50**	**290**	**6**	**.04**	**.03**	**.40**	**5**
182	1.2	82	223	161	.06	.17	8.49	0
102	1.5	4	477	5	.15	.09	1.50	ND
10	.4	19	93	2220	.02	.04	.24	10
103	.6	17	249	217	.22	.07	1.61	6
66	**3.1**	**76**	**397**	**14000**	**.19**	**.26**	**ND**	**35**
63	**2.4**	**4**	**390**	**4000**	**.04**	**.10**	**.50**	**48**
172	1.2	124	126	560	.07	.51	.06	0
153	**1.7**	**17**	**401**	**0**	**.20**	**.11**	**.70**	**31**
10	**.3**	**2**	**191**	**100**	**.09**	**.57**	**.30**	**4**
28	.9	23	228	7400	.05	.07	.50	41
18	**.6**	**ND**	**ND**	**ND**	**ND**	**ND**	**ND**	**ND**
72	**1.2**	**0**	**0**	**11600**	**.16**	**.44**	**1.20**	**80**
45	1.1	8	290	2600	.10	.10	.50	24
20	.5	9	150	330	.05	.03	.19	4
2	**1.2**	**9**	**204**	**ND**	**.01**	**.01**	.03	0
93	.1	49	152	126	.04	.16	.08	1
ND	**ND**	**ND**	**ND**	**ND**	**ND**	**ND**	**ND**	**ND**
104	**1.2**	**4**	**370**	**0**	**.10**	**.45**	**4.11**	**4**
41	**.7**	**16**	**202**	**3031**	**.04**	**.06**	**.43**	**25**
79	**.79**	**0**	**539**	**ND**	**.11**	**.12**	**1.83**	**0**
29	**.3**	**2**	**55**	**0**	**.06**	**.01**	**.10**	**8**
33	**1.9**	**2**	**257**	**5000**	**.07**	**.14**	**.20**	**45**
14	.1	0	181	205	.09	.04	.28	53

	Source	Energy (kcal)	Prot (g)	Fats (g)	Carb (g)	Calc (mg)
Pears, raw	6	59	.4	.4	15	11
Pears, Prickly, raw with seeds	**3**	**60**	**1.4**	**1.4**	**12**	**46**
Peas, raw	1	81	5.4	.4	14	25
Peanuts with skin	5	567	25.7	49.1	16	58
Peppergrass	**2**	**32**	**2.6**	**.7**	**ND**	**81**
Pecans, dry	5	667	7.8	67.6	18	36
Persimmons, native, raw	**6**	**127**	**.8**	**.4**	**34**	**27**
Pokeweed, shoots, boiled	**1**	**20**	**2.3**	**.4**	**3**	**53**
Potatoes, baked w/skin	1	109	2.3	.1	25	10
Potatoes, sweet, baked	1	103	1.7	.1	24	28
Purslane, raw	**1**	**16**	**1.3**	**.1**	**3**	**65**
Raspberries, raw	6	91	.8	.1	23	11
Rice, brown, cooked	8	112	2.3	.8	24	10
Sheppard's Purse	**2**	**33**	**4.2**	**.5**	**ND**	**208**
Sorrel, Sheep	**2**	**77**	**1.9**	**ND**	**ND**	**55**
Sorrel, Wood	**2**	**ND**	**.1**	**ND**	**ND**	**ND**
Soybean, green, cooked	1	141	12.3	6.4	11	145
Spinach, raw	1	22	2.9	.4	4	99
Squash, summer, raw	1	20	1.2	.2	4	20
Squash, winter	1	37	1.5	.2	9	31
Strawberries, Wild, raw	**2**	**37**	**.7**	**.5**	**ND**	**21**
Sunflower seeds, dried	5	570	22.8	49.6	18	116
Thistle, Sow	**2**	**20**	**2.4**	**.3**	**ND**	**93**
Tomato, raw	1	19	.9	.2	7	7
Turnips	1	18	.7	.1	5	22
Turnip Greens, cooked	1	20	1.4	.2	4	137
Violet, leaves	**2**	**ND**	**ND**	**ND**	**ND**	**ND**
Walnuts, black, dried	**5**	**607**	**24.3**	**56.5**	**12**	**58**
Wheat, sprouted	8	198	7.5	1.3	42	28

1 Composition of Foods - USDA - Vegetables - 8-11 1984
2 Mother Earth News #6 Robert Shosteck
3 Foods & Nutrition Encyclopedia - Volume 1, Ensminger 1983 Pegus Press.
4 Composition of Foods - USDA - Beef Products - 8-13 1986
5 Composition of Foods - USDA - Nut & Seed Products - 8-12 1984
6 Composition of Foods - USDA - Fruits & Fruit Juices - 8-9 1982
8 Composition of Foods - USDA - Cereal Grains - 8-20 1989
7 Composition of Foods - USDA - Poultry Products - 8-5 1972

Phos (mg)	Iron (mg)	Sod (mg)	Pot (mg)	A (IU)	Thia (mg)	Ribo (mg)	Niac (mg)	Asor (mg)
11	.2	0	125	20	.02	.04	.10	4
32	**1.2**	**ND**	**ND**	**10**	**.02**	**.03**	**.40**	**22**
108	1.5	5	244	640	.27	.13	2.90	40
383	3.2	16	717	0	.66	.13	14.15	0
76	**1.3**	**14**	**606**	**9300**	**.08**	**.26**	**1.00**	**69**
291	2.1	1	392	128	.85	.13	.89	2
26	**2.5**	**1**	**310**	**ND**	**ND**	**ND**	**ND**	**66**
33	**1.2**	**ND**	**ND**	**8700**	.07	.25	1.10	82
57	1.4	8	418	ND	.11	.02	1.39	.1
55	.5	10	348	21800	.07	.13	.60	25
44	**2.0**	**45**	**494**	**1320**	**.05**	**.11**	**.48**	**21**
9	.4	3	94	33	.02	.03	.44	9
77	.5	1	79	ND	.10	.01	1.33	0
86	**4.8**	**ND**	**394**	**1554**	**.25**	**.17**	**.40**	**36**
82	**5.0**	**ND**	**ND**	**ND**	**ND**	**ND**	**ND**	**ND**
ND	**ND**	**ND**	**ND**	**2800**	**ND**	**ND**	**ND**	**ND**
158	2.5	ND	ND	156	.26	.16	1.3	17
49	2.7	79	558	6715	.08	.19	.72	28
35	.5	2	195	196	.06	.04	.55	15
32	.6	4	350	4060	.10	.03	.80	12
21	**1.0**	**1**	**164**	**60**	**.03**	**.07**	**.60**	**59**
705	6.8	3	689	50	2.29	.25	4.50	2
35	**3.1**	**ND**	**ND**	**2185**	**.70**	**.12**	**.40**	**5**
207	.4	8	207	1133	.06	.05	.60	17
19	.2	50	135	0	.03	.02	.30	11
29	.8	29	203	5498	.05	.07	.41	27
ND	**ND**	**ND**	**ND**	**8200**	**ND**	**ND**	**ND**	**210**
464	**3.1**	**1**	**524**	**296**	**.22**	**.11**	**.69**	**ND**
200	2.4	16	169	ND	.22	.16	3.09	3

9 Composition of Foods - USDA - Dairy & Egg Products - 8-1 1976
10 Composition of Foods - USDA - Snacks & Sweets - 8-19 1991

ND - No Data. Note: 100 milligrams (mg.) equals about 3 1/2 ounces. For converting IU (International Units) to RE (Retinol Equivalents), divide the International Units by 10 based on the ratio used in the USDA publication "Composition of Foods".

APPENDIX D

A PERSONAL NOTE FROM KEN

Many people increase their faith in God during difficult times. Those of us who have been in war refer to it as "foxhole religion" or "There are no atheists in foxholes."

It was in Viet Nam that my testing began. The duty days were fourteen hours long, seven days a week. There was just enough time to finish guard mount, eat, wash if there was water, and go to sleep. It seemed that before I was even settled in bed, it was morning and I awoke while laying in a puddle of sweat from the hot, humid night before. Many nights I would be awakened by several rapid firings of a 40 millimeter cannon next to my position, usually a firefight caused by something tripping a warning flare or a claymore mine.

And so it went, day in and day out causing the stress to build. Some cannot deal with the separation from loved ones and the constant fear for their lives. Early in my tour, our "point man" armed with a twelve gauge shotgun "cracked" and proceeded to shoot up the compound. Later we had a Captain who, after six months in country, went on R&R (Rest & Recuperation) to meet his wife in Hawaii. When he got off the plane, she handed him divorce papers. The night he returned to his unit he was found with a 45 caliber pistol stuffed in his belt and a knife in his hand trying to attack his bunkmate.

In my case, I had an emotional release outlet. About once a week, as I looked down the valley we were protecting, I would hear the ever present gunfire. It was usually punctuated with mortar or tracked anti-aircraft guns we used on enemy ground forces. As I surveyed my fear, frustration, and hopelessness, I would become overwhelmed and sob for what seemed like forever. Then composure would surface and I would be OK for several days.

If only I could survive this tour and return to the states! With the survival skills and independence I had so intensely acquired, I could deal with anything. Fortunately, my life was spared and I returned home.

Working as a District Marketing Manager with Ford Motor Company, covering Maryland, Delaware and Washington, DC., I was on the "fast track" leading to the "Crystal Palace" in Detroit. I had a nice home, an executive position, and numerous material possessions, but I was dissatisfied with my work, other people, and life in general. So I tried climbing one step higher on the corporate ladder and accepted a position with FMC as a Regional Manager over 13 states. This meant a marketing staff, a larger title, more money, increased travel, and deeper frustration.

My disillusion bottomed-out one day while I was on an automobile trip from Baltimore to Pittsburgh. I felt compelled to review my life since college. With a degree in Industrial Design and an MBA from Auburn University, I had been a design engineer with Lockheed; a business consultant; a marketing manager with Ford Motor Company; and now I was a regional manager with FMC, a large Fortune 100 company. I was supposed to be thankful for my professional growth and lifestyle, but, instead, I felt empty.

Tracing my path, regardless of the increased income and assets I acquired, I sensed there was still something I needed to feel successful and satisfied. I had finally talked my way into a job in which I knew I was not meeting the performance criteria, and there were problems in my marriage. The reality of losing control over my life became more depressing. I knew no real happiness and could not achieve any sense of peace in my life.

My former assumption that happiness is related to success had been quickly altered one day on a plane flight as I sat beside the captain of a pleasure-boat belonging to a very wealthy businessman–so successful that his pleasure boat was 150 feet long and had a full time crew of three. Such a craft symbolized everything I had been striving for in my life. I was talking to someone who was next to that wealth and success, who could tell me first hand about what it was really like at the top. After we talked for a while, I asked him: "Your employer is very successful; what is it like at the top?"

The man was thoughtful for a minute before he said: "Let me answer it this way. Do you know what my employer does on Friday night? He comes down to the boat, and we sit around and get drunk together." There was a profound silence while my dreams crumbled–it was unnerving to realize that success did not guarantee happiness.

Later on that day as I was driving toward Pittsburgh, my considerations were halted by a roadblock on the turnpike: a blizzard had shut off all access to the Pittsburgh area. As I started back to Baltimore, hopelessness overwhelmed me. I concluded that life was not worth the effort, and I needed to "punch out." It seemed that my only solution was to commit suicide. Exploring ways to do it, I decided to hit one of the bridge supports. Then I thought of the direct possibility, with the way my luck was going, that I could not do it right and I would only be crippled.

I had gone to church all my life and was basically a good person, but here I was, brooding in deepest despair. Suddenly the voice of a radio preacher broke through my desolation, and I heard him say, "If you have reached the top and find there is nothing there; if you do not have peace in your life, try God." It seemed he was speaking directly to me! The preacher said: "Put Jesus to the test and see if he can straighten out your life."

I knew I didn't have anything left to lose so I pulled off the turnpike and gave myself up to following the man's clear instructions. I remembered John 3:16. Then I prayed: "Jesus, if you can straighten me out, then I need you in my life. I believe you died for my sins, and you can give me everlasting life. I accept you as my personal savior." Then it happened! My letting go of my life in a simple act of faith caused a peace to flow over me that I can't explain. I felt as if a two-hundred pound backpack was lifted off my shoulders.

Since that day, my life has had its difficulties as well as good times, but through them, God has provided me a "peace that passes all understanding." I have no more fears nor worries because I know that no matter what happens, now I have a special Presence supporting and guiding me. Not only did I receive inner healing and eternal salvation; my relationship with my family also began to be wonderful.

That small town radio preacher in Pennsylvania will never know the impact his simple message had on my life, my family, and my two children! Now I want my simple message to help others.

As we see the present foundations of our world crumbling, we can find peace and strength. Remember Philippines 4:6, 7 says: "Do not be anxious about anything, but in everything through prayer and petition, with thanksgiving, make your requests to God. And the peace of God, which transcends all understanding, will guard your hearts and minds in Christ Jesus". Someone who has promised never to leave or forsake us–Rest in peace; God is awake.

Ken Larson

INDEX